Earth Resources

Earth Resources

Brian J. Skinner

Yale University

Prentice-Hall, Inc.,   Englewood Cliffs, New Jersey

For Catherine, Adrienne, Stephanie, and Thalassa

Design by Walter Behnke

Illustrations by Felix Cooper

PRENTICE-HALL INTERNATIONAL, INC., *London*

PRENTICE-HALL OF AUSTRALIA, PTY., LTD., *Sydney*

PRENTICE-HALL OF CANADA, LTD., *Toronto*

PRENTICE-HALL OF INDIA PVT. LTD., *New Delhi*

PRENTICE-HALL OF JAPAN, INC., *Tokyo*

Current printing (last digit):
10 9 8 7 6 5 4 3 2 1

FOUNDATIONS OF EARTH SCIENCE SERIES

A. Lee McAlester, Editor

(p) 13-222653-7

(c) 13-222661-8

Foundations

of Earth Science Series

Elementary Earth Science textbooks have too long reflected
mere traditions in teaching rather than the triumphs and
uncertainties of present-day science. In geology, the time-
honored textbook emphasis on geomorphic processes and
descriptive stratigraphy, a pattern begun by James Dwight
Dana over a century ago, is increasingly anachronistic in an
age of shifting research frontiers and disappearing boundaries
between long-established disciplines. At the same time, the
extraordinary expansions in exploration of the oceans, atmosphere, and interplanetary space within the past decade have
made obsolete the unnatural separation of the "solid Earth"
science of geology from the "fluid Earth" sciences of oceanography, meteorology, and planetary astronomy, and have
emphasized the need for authoritative introductory textbooks
in these vigorous subjects.

Stemming from the conviction that beginning students de-
serve to share in the excitement of modern research, the
Foundations of Earth Science Series has been planned to provide brief, readable, up-to-date introductions to all aspects of
modern Earth science. Each volume has been written by an

authority on the subject covered, thus insuring a first-hand treatment seldom found in introductory textbooks. Four of the volumes—*Structure of the Earth, Earth Materials, The Surface of the Earth,* and *Earth Resources*—cover topics traditionally taught in physical geology courses. Three volumes—*Geologic Time, Ancient Environments,* and *The History of Life*—treat historical topics, and the remaining three volumes—*Oceans, The Atmosphere,* and *The Solar System*—deal with the "fluid Earth sciences" of oceanography, meteorology, and astronomy. Each volume, however, is complete in itself and can be combined with other volumes in any sequence, thus allowing the teacher great flexibility in course arrangement. In addition, these compact and inexpensive volumes can be used individually to supplement and enrich other introductory textbooks.

Acknowledgements

Grateful acknowledgement and thanks are due to Mrs. Evelyn Burn, Mrs. Heather Bogarty, and Mrs. Jan Paplham for uncomplainingly effecting the transition from the illegibility of script to the legibility of typed manuscript, and to Mr. Arthur Vergara of Prentice-Hall, Inc., for his efforts to add readability to the legibility.

Contents

Introduction

All forms of life draw resources from the Earth. Man alone has
systematically used the available materials to shape the unique
form of controlled living we call civilization. In so doing, he
has learned to combat climatic extremes and to increase vastly
the Earth's yield of palatable foods above Nature's random
growth. As a consequence, we have expanded our occupation
of the globe to its farthest reaches and have proliferated our
species far beyond the numbers capable of living in equilibrium with an uncontrolled Nature. Maintenance of Earth's
huge population now depends directly on a continuing supply
of the resources needed to fuel and operate civilized society.

But is there sufficient for a healthy future, and are the vital
resources sufficiently accessible to allow easy exploitation?
Empires have flourished repeatedly through history because
of their control over rich and easily exploited mineral resources, but they have withered with the same frequency as
these riches expired. Are the world's remaining resources so
distributed that the historical pattern of power dependence
on resource availability is now a thing of the past, or is it still
the key to the future?

1

Study of the abundance and distribution of the Earth's resources is, in its more basic aspects, a branch of geology, and this book very properly belongs in an Earth Science Series. It is also the starting point for any study of civilized history and of man's occupancy of the Earth. It cannot be ignored in any field of learning, but it is essential in any predictive considerations of the future, for all of us daily become more dependent on maintenance and orderly change in the civilization around us.

1

What constitutes

an earth resource?

Our entire society rests upon—and is dependent upon—our water, our land, our forests, and our minerals. How we use these resources influences our health, security, economy, and well-being. (John F. Kennedy, Message on National Resources; Congress, February 23, 1961.)

We are all familiar with the term *natural resources;* it refers to the supplies of food, building and clothing materials, minerals, water, and energy that we draw from a bountiful Earth and that we need to sustain life and our complex civilization. We are not so familiar with the concept that some resources, such as food, can be replaced each year by the seasonal growth of plants and are thus continuously *renewable resources* while others, like minerals, coal, and oil, are *depleting* or *nonrenewable resources* because the Earth contains fixed supplies which are being steadily consumed.

The expansion of population raises two central questions, the answers to which provide the key to the long-term future of man. The first question is about life itself—at what maximum *rate* can the renewable food resources be produced without irretrievably eroding and ruining the surface of the

Earth? The answer, when known, will provide the key to how many people the Earth can sustain on a continuous basis. The second question is about civilization—does the Earth contain enough of the fuels, metals, building materials, fertilizers, and other depleting resources to support and expand our increasingly complex civilization?

Whether we would ever wish to live in a world so populous that every resource was strained to the maximum rate of production is very doubtful; the attempt would in any case be foolhardy, because it would leave no margin for error. Nevertheless, it is essential that we try to understand the limits of our natural resources, because we are already approaching the exhaustion point of present supplies of such scarce resources as helium gas and mercury.

Renewable vs. Nonrenewable Resources

Food, natural fibers like cotton, and forest products such as lumber and paper pulp are derived from plants, and because plants are seasonally regenerated, all the resources that they supply are renewable. The growing plant removes carbon dioxide from the atmosphere and by the process of photosynthesis uses the Sun's energy to combine the carbon dioxide with water from the soil to make carbohydrates and oxygen which it releases to the atmosphere (Fig. 1–1). Plant carbohydrates are the basic food supply for all animals, including man, either by direct consumption or indirectly by consumption of plant-eating animals. The consumption of plants by animals is essentially a decay process in which carbohydrates are again reduced to water and carbon dioxide by oxygen from the atmosphere; thus, the chemical reason

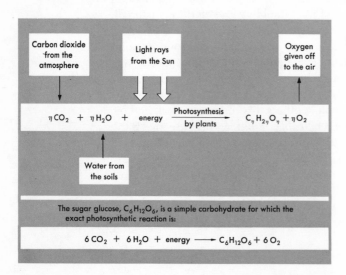

FIGURE 1-1   *Photosynthesis produces carbohydrates, our basic food resource.*

why animals breathe oxygen is to burn their carbohydrate food supply. The energy trapped by the growing plant during photosynthesis is released in the decay process and provides the energy needed for animal life. The growth and consumption process is cyclic; similarly, the same carbon dioxide and water can be recycled endlessly. The only component of the cycle that is not returned to its original state is the Sun's energy, and this is limitlessly abundant. Provided the soils and waters that nurture our crops are given adequate care, the supply of food, fiber, and forest products that we draw from them need not be exhausted.

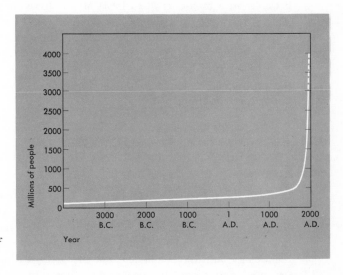

FIGURE 1–2 *Growth rate of the human population.*

Although production of food resources can be endless, there is a limit to the annual *rate* of production. Plants grow only in the narrow zone where the atmosphere meets the surface of the Earth—the zone where the essential ingredients, carbon dioxide, water, and light energy are all readily available in the right proportions. The Earth's surface has a fixed area, and the rate at which the Sun's energy falls on it is constant, so that these two fixed quantities limit the ultimate rate at which plant food can be produced. Though we may learn to exploit the surface more effectively and may even develop plants that use the Sun's energy more efficiently than present plants, limits fixed by area and amount of energy will remain.

The population of the Earth has increased continuously since man's first appearance (Fig. 1–2). It is now increasing at a much faster rate than our increase in plant food. Even with enormous efforts to expand food production from all sources, including the oceans, it is inevitable that the population must cease growing at some point or else the food consumption rate will exceed the production rate. A recent critical study of resources and man sponsored by the National Academy of Sciences, estimated the Earth's theoretical carrying capacity at approximately 33 billion people—10 times the world's present population. A measure of the severity of the renewable resource crisis is that even if all possible means to increase the food productivity rate were used and strict regimentation of diet were practised, the 33 billion people could only be fed at a level of chronic near-starvation. Furthermore, the food production rate would be a theoretical maximum, attainable for a short spurt, but not possible on a long-continued basis of hundreds of years. If we maintain the present birthrate, this doleful day will be reached about 100 years from now. Clearly, we should be foolish to reach such a disaster point, and a more practical population limit, by which we can live in equilibrium with a stable rate of food production, must be reached. The definition of this practical limit, and the means of holding it in stable balance, are two of the greatest problems facing mankind today.

Nonrenewable Resources

Even the present population level on Earth could not have been reached without the incredibly complex system of health control, power distribution, transportation, and communication that underlies our civilization. But civilization is totally dependent on the minerals and mineral fuels needed for complex technology, and these, unlike plant products, are not formed by rapid cyclic processes and cannot be regenerated once they have been mined or pumped from the Earth. The Earth contains fixed amounts of all the nonrenewable resources and most such resources can be used only once. Coal and oil once burned, phosphatic fertilizers once distributed, and clays once fired into bricks are permanently consumed. Even metals like iron, lead, and copper, which can be recovered in scrap form, are only partially recovered after each cycle of use and in the long run are also permanently consumed. A few of the Earth's nonrenewable resources, and most importantly water, are not permanently consumed. We may locally redistribute water by man-made reticulation and pumping systems, but our control over it is temporary and it soon returns to the evaporation-transpiration-precipitation cycle that governed its original distribution. No water is lost in the continuous cycle, and in contrast to the depleting resources, water is a *reusable resource*. It is important to realize that we are discussing water on a global scale, because water pumped from a lake or underground supply at a rate faster than natural replenishment would certainly be a case of depleting local resources. Provided we do not irretrievably pollute the waters of the Earth by needlessly distributing the debris from our usage of the depleting resources, we should be able to reuse water endlessly.

This book is about the nonrenewable resources, both depleting and reusable, and is largely geological in content, because geology is the science that deals with their distribution.

The Increasing Consumption of Nonrenewable Resources

Man is distinguished from the other animals by his abilty to use the materials of the Earth to feed, clothe, and warm himself, to communicate with other men, and to transport himself at will. Among the earliest evidences of emerging man are the crude stone and bone implements he utilized at the dawn of civilization; as his sophistication and mastery over his environment increased, he used an ever-expanding range of materials—copper, bronze, iron, clays for pottery and dwellings, glass, pigments, cements, and abrasives. Each forward step in man's history brought demands for new materials and for ever-increasing quantities.

The dramatic increase in the production rate of mineral products in the United States can clearly be seen from figures published by the U.S. Bureau of Mines (Fig. 1–3). The total dollar value of the mineral fuels—coal, oil, and natural gas—as well as of the metals such as copper, lead, zinc, and iron, and of the nonmetallic minerals such as clays, cements, salts, and fertilizers pro-

FIGURE 1–3   *Total value of minerals produced annually in the United States. (After U.S. Bureau of Mines.)*

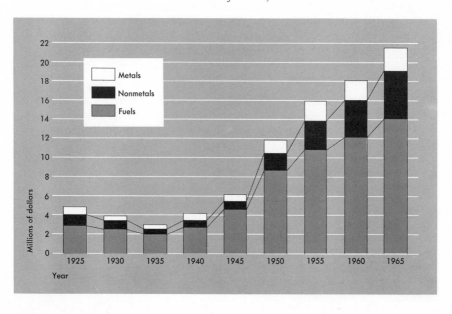

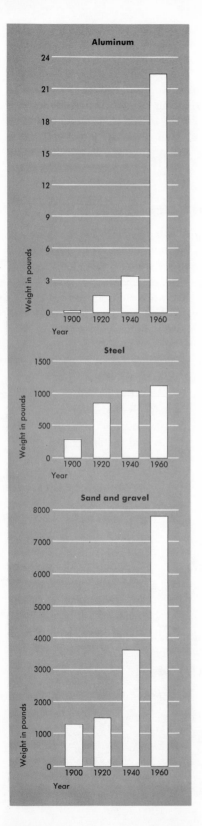

duced in the United States rose from a billion dollars a year in 1900 to 22 billion in 1965, an increase of 22 times, while the population increased from 76 million to 195 million, an increase of only 2.6 times.

Impressive though this growth rate is, it does not reveal the whole story. Concealed in the numbers from which Fig. 1–3 is prepared are other factors such as the decreasing purchasing power of the dollar. Furthermore, the curve does not take into account all the mineral products imported into the country to supplement local production. A more definitive question to ask would be: At what rates are mineral products consumed within the country? Fortunately, this question can also be answered from data compiled annually by the U.S. Bureau of Mines. When we examine the consumption figures, we find two components to the growing consumption rate. The first component is the increasing population—more people need products to support them. The second component is an index of the increasingly complex technological underpinning of our society. If we divide the annual gross consumption of a mineral product by a number equal to the population, we get the annual per capita consumption. For almost all mineral products in an industrial country such as the United States, the per capita consumption is steadily increasing (Fig. 1–4).

The rate of increased consumption varies among the minerals, as is demonstrated by the dramatic increase in the per capita consumption of aluminum compared to the more gradual increase in that of steel. Nor do all the consumption rates increase steadily year by

FIGURE 1–4 *Increasing annual per capita consumption of some major mineral products in the United States expressed in pounds. (After U.S. Bureau of Mines.)*

What constitutes an earth resource?

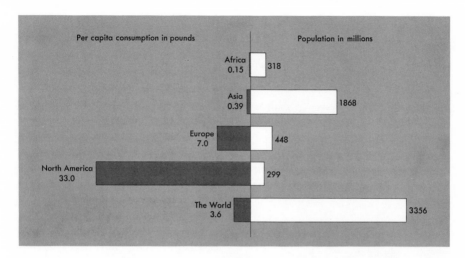

FIGURE 1-5 *Per capita consumption of aluminum in different continents, 1964. (After S. Brubaker, 1967. No data available for South America.)*

year, because short-term fluctuations occur, such as those caused by wars and depressions. The combined effects of increasing per capita consumption and increasing population have raised the gross mineral consumption rate to such an extent that from 1930 to 1960 the people of the United States alone used more mineral products than all the previous peoples of the world had collectively used since the beginning of history. This relentless growth rate shows every sign of continuing to increase. Data gathered by the U.S. Bureau of Mines show that by 1965, annual per capita consumption in the United States was: steel, 1,350 pounds; sand and gravel, 8,700 pounds; sulfur, 82 pounds; copper, 20 pounds; aluminum, 35 pounds; crude oil, 700 gallons; coal, 4,850 pounds; and helium gas, 3.9 cubic feet.

Because industrialization is very unevenly developed around the world, the geographic consumption of mineral resources is not uniform. As an example, consider the very uneven per capita consumption of aluminum in the year 1964 (Fig. 1–5), and note the low consumption rate in Asia where half the world's population resides. Simply to raise the world's per capita consumption of all mineral resources to the present rate enjoyed by the population of the United States would severely tax the known reserves of many metals. If we are confidently to face both an increasing per capita consumption *and* an increasing population growth rate, it is imperative that we appreciate and understand the nature of our present and potential resources.

Types of Mineral Resources

Throughout this book we shall be discussing mineral resources in their most general sense, and the term will be taken to include all nonliving, naturally occurring substances that are useful to man whether they are inorganic or organic. Thus all natural crystalline solids, fossil fuels such as petroleum and natural gas, as well as the waters of the earth and gases of the atmosphere fall under this definition of mineral resources.

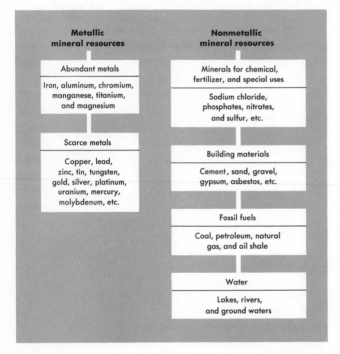

FIGURE 1-6 *Types of mineral resources.*

The different types of mineral resources can be classified on the basis of use (Fig. 1-6). First, there are the chemical elements which are useful because of their metallic properties. These can be further subdivided into two groups on the basis of their abundance—those that are relatively abundant on the Earth's crust, like iron, aluminum, manganese, and titanium, and those that are relatively scarce, like copper, lead, and zinc. Secondly there are the non-metallic materials which can be subdivided into four groups on the basis of their uses:

1. The nonmetallic materials, such as sodium chloride, calcium phosphate, and sulfur, of primary use to the chemical and fertilizer industries.
2. The nonmetallic materials, such as sand, gravel, crushed stone, gypsum, and cements, used primarily as building and construction materials.
3. Fuels to meet our growing energy needs, but particularly the all-important remains of plant and animal matter—the fossil fuels, such as coal, petroleum, and natural gas, on which we currently rely so heavily for our energy supplies.
4. Water, the single most important mineral resource of all, without which we could neither grow our food nor keep our technology operating.

In the chapters that follow, the different mineral resources will be discussed following the classification in Fig. 1–6.

Even though we have classified mineral resources on the basis of use, we must remember that they are found as chemical compounds and that the processes which preferentially concentrate elements and compounds in the Earth are chemical processes that are not necessarily controlled in any way by man's use of the end product. Before we proceed to discuss specific mineral resources, we shall briefly consider the chemical elements found in the Earth and, in the next chapter, their abundance, distribution, and availability.

The Elements We Use

The chemical elements can be arranged in a periodic order (Fig. 1–7) based on their atomic masses and on the distribution of electrons in the electronic shells surrounding their nuclei. There are now 103 elements known, and of these, 88 have been found on Earth. The heaviest of the natural elements, uranium, is number 92 in the periodic table. All the elements above 92 are man-made and are collectively called the transuranic elements; they are all radioactive and relatively short-lived. The four elements lighter than uranium that have not been found occurring naturally on Earth—technetium, promethium, astatine, and francium (elements number 43, 61, 85, and 87 respectively)—are also radioactive and short-lived. All of the 88 natural elements have some practical value for man, although the uses of some are so specialized and the demand for them so small that we can hardly consider them essential.

Each year the U.S. Bureau of Mines publishes a very informative volume titled *Minerals Yearbook*, in which the annual world consumption of all mineral products is estimated. From these data we can calculate the consumption

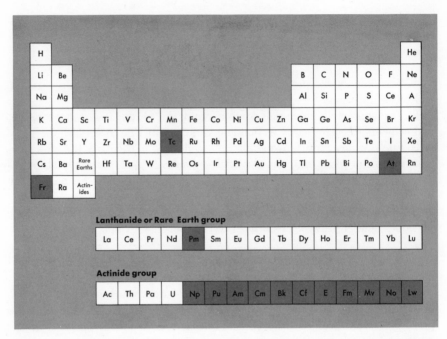

FIGURE 1-7 *Periodic table of the elements. Elements in each column have similar chemical properties. Elements in each row increase in mass from left to right. Rare Earth elements and actinide elements have very similar properties and appear as single entries in the table. Shaded entries are elements not known to occur naturally on Earth. Names of the elements, and their symbols, are listed in Appendix, Table 1.*

rates of the principal elements, regardless of the form or compound in which they are used. If we exclude sand, gravel, and crushed stone (because they are nonspecific groupings of elements) and consider the 15 Rare Earth elements as a single entry (because of their similarity in properties and uses), there are 48 elements that are annually consumed in amounts greater than 100 tons (Fig. 1–8).

The element consumed in greatest quantities is carbon, a reflection of our reliance on carbon-rich fossil fuels such as crude oil, coal, and natural gas as energy sources. Carbon is followed by sodium, used principally in the form of sodium chloride for human consumption and as a raw material for the chemical industry, and by iron, the major structural metal of our civilization. It may be surprising to find that four of the next five most heavily consumed elements are used as fertilizers and soil conditioners. Nitrogen is used in nitrates and ammonia compounds, sulfur in ammonium sulfate and superphosphate, potassium in potassium chloride, and calcium in crushed limestone

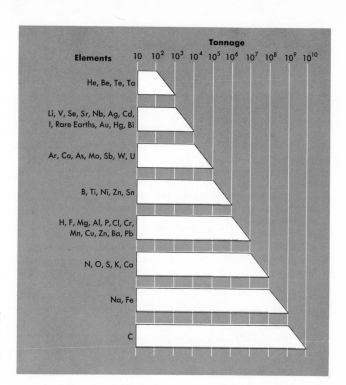

FIGURE 1-8 *Annual world consumption of elements in tons, excluding sand, gravel, and stone. (After U.S. Bureau of Mines.)*

and burned lime. Each of these elements, of course, has other highly important uses—calcium, for instance, is consumed in huge quantities in the production of cements—but it is as fertilizers for the production of adequate food supplies that they are most important. It is not until we reach production levels of less than 10 million tons per year that we come upon such metals as copper, lead, and zinc, and the other elements that we commonly think of when mineral resources are discussed.

Estimation of Resources

The word *resource,* in the context in which we shall use it, means a supply, and, as with all supplies, we are necessarily concerned with how much we have available and can depend on for the future. In attempting to estimate the world's reserves of the many resources we now use, we must make a number of highly subjective guesses. For example, in projecting estimates into the future, we must guess at the technological advances that might make presently uneconomic deposits workable; and since in estimating how much ore is present in a given deposit we are faced with inability to see and measure the underground limits of the ore except by very costly mining and drilling procedures, geological guesswork is used instead. Depending on the amount of guesswork involved, we can separate reserve estimates into categories of increasing probability of their being correct. The grouping most convenient for our purposes was first developed by geologists of the United States Department of the Interior.

The starting place is the point of highest probability, the mineral deposit for which we have all the information necessary to say exactly how much ore is present and exactly how rich it is—in this case we have a *measured reserve*. If the measurements were not taken as closely as we might wish, some geological guesswork would be involved in making the estimate, and we would only be justified in calling our estimate an *indicated reserve*. If the estimate were based largely or entirely on geological inference and experience, the probability of being correct on all counts would be low, so the estimate would only be an *inferred reserve*. These terms of decreasing certainty refer only to resources that are economically profitable to work. Whether a mineral deposit falls in one of the reserve categories depends on many things besides its size and richness, principally the relative concentration of the material and accessibility to markets, but including such things as political boundaries, taxation laws, and price subsidies. An example of the latter in recent years was the development of uranium reserves during the early 1950's when strategic needs led the U.S. government to support the industry by long-term market contracts. As these contracts expired and the mines had to sell their produce on the open market, lower prices forced many to become unprofitable. Clearly, the uranium is still present in the ground and can be mined when economic conditions warrant its recovery. To cover this category of identifiable concentrations of materials that may become economically recoverable in the future, we will use the term *potential resources*, introduced in this context by P. T. Flawn. The certainty with which we can estimate the magnitude of the potential resources decreases the further we move from common experience, the more unconventional the materials we include in the estimate, and the further we project our guesses of economic and technological changes into the future. The one point that seems clear, however, is that man will always mine selectively, seeking out the richest remaining ores first. The suggestion occasionally made that man will eventually mine common rocks, essentially ignoring the local concentrations we commonly call ore deposits, is not likely to be realized.

2

Abundance and availability
of the earth's resources

But the needed materials which can be recovered by known methods at reasonable cost from the earth's crust are limited, whereas their rates of exploitation and use obviously are not. (Walter R. Hibbard, Jr., in "Mineral Resources: Challenge or Threat?," Science, v. 160, p. 143, 1968.)

Man has learned to draw his resources from all accessible portions of the Earth, including the oceans and the atmosphere. The present and projected exploitation of the Earth's resources, in addition to the scattering of debris from the usage, is so great that we must attempt to look beyond geographic boundaries and local environments to get a realistic view of the total supply. In order to place this supply in correct perspective, we should first look at the essential features of the planet we live on.

The Earth

The Earth has a mass of 6.5×10^{21} tons and is composed of 88 different elements; the total tonnage of any element is truly enormous, even for those present at extremely low concentration levels. Although we cannot sample interior regions directly, we can estimate their compositions from several lines of investigation.

There is only a restricted group of elements, with low to intermediate atomic masses, that can account for the mass and density of the Earth, and accordingly these elements must be the most abundant ones. This statement is corroborated by other evidence, such as the compositions of volcanic rocks erupted from the Earth's deeper portions, and meteorites, believed to have come from a former planetary body with a composition similar to that of the Earth. All the evidence indicates that ten elements account for 98.8 per cent of the entire Earth mass. The average composition shown in Fig. 2–1 applies to the

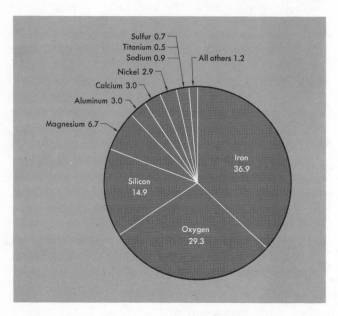

FIGURE 2–1 *Abundance of major chemical elements in the Earth. (After E. Niggli, 1928.)*

Earth as a whole, yet the Earth is not a homogeneous body. We know from direct observation and from such evidence as the paths followed by seismic waves through the Earth that it contains a series of concentric shells of vastly differing composition and density (Fig. 2–2). At the center is a metallic *core* consisting predominantly of iron and nickel, surrounded by a *mantle* of dense oxide minerals rich in iron and magnesium; the core and mantle together account for more than 99.6 per cent of the total mass of the Earth. Above the mantle is the Earth's *crust,* which is the only portion of the solid Earth we actually observe, and which accounts for 0.375 per cent of the Earth's mass.

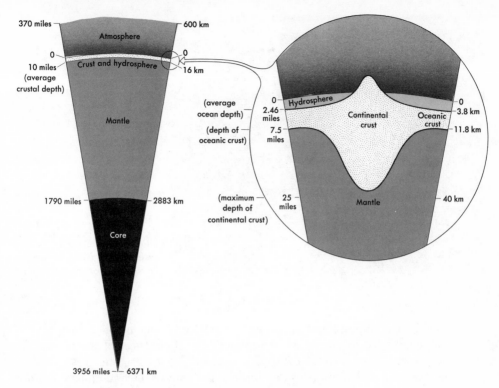

FIGURE 2–2   *Concentric structure of the Earth zones.*

It is of two parts, one that projects above the oceans and one that lies below; the portion above, in addition to a narrow sea-covered fringe around each continent, is called the *continental* crust; the portion below the oceans is the *oceanic* crust.

At the Earth's surface are the oceans, lakes, and rivers that, together with the water trapped in holes and fractures in soil and near-surface rocks, are called the *hydrosphere,* which accounts for 0.025 per cent of the Earth's mass. Enclosing everything is the gaseous envelope of the *atmosphere,* which accounts for only 0.0001 per cent of the mass. It is from the three outermost zones—the crust, hydrosphere, and atmosphere, which together account for only 0.4 per cent of the total mass of the Earth—that we draw our present resources and to which we must look for those of the future. The mantle and the core are so inaccessible that they cannot be seriously considered as potential sources of minerals.

The Atmosphere

The atmosphere is continuously mixed because it is gaseous and accordingly has an essentially uniform composition; it is accessible and therefore easy to

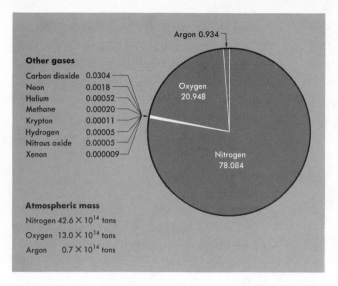

Other gases

Carbon dioxide	0.0304
Neon	0.0018
Helium	0.00052
Methane	0.00020
Krypton	0.00011
Hydrogen	0.00005
Nitrous oxide	0.00005
Xenon	0.000009

Argon 0.934

Oxygen 20.948

Nitrogen 78.084

Atmospheric mass

Nitrogen 42.6×10^{14} tons

Oxygen 13.0×10^{14} tons

Argon 0.7×10^{14} tons

FIGURE 2–3 *Average composition of clean dry air on a volume percentage basis.*

sample. The atmospheric composition is relatively simple and is known with considerable accuracy, so our estimates of the abundances of the various atmospheric gases fall in the category of measured reserves. Three gases—nitrogen, oxygen, and argon—account for 99.9 per cent of the atmospheric volume (Fig. 2–3), with nitrogen, a basic plant fertilizer component, as its most important constituent (see Chapter 5).

Oxygen and argon, together with the rarer gases neon, xenon, and krypton, are also recovered from the atmosphere, but in relatively small amounts. The uses to which gases are put do not permanently remove them from the atmosphere; thus, we can classify them as reusable resources. Also, because of the small amounts used, their temporary removal has no observable effect on the atmospheric composition.

For the few recoverable elements concentrated in it, the atmosphere provides an essentially limitless source.

The Hydrosphere

The hydrosphere is the body of condensed water on and near the surface of the Earth, and water itself is its most vital resource. Owing to its extreme importance, water will be discussed separately in Chapter 8. The hydrosphere is the site of other vital resources: The oceans, which cover 70.8 per cent of the Earth's surface to an average depth of 2.46 miles, act as a collection reservoir for many of the soluble materials formed on Earth, such as those released from rocks and soils by weathering and those given off in volcanic gases. Through the ages, they have reached their present composition level of 3.5 per cent dissolved solids by weight in solution.[1] A dissolved solid content of 3.5 per cent amounts to approximately 160×10^6 tons of dissolved matter in every cubic mile of sea water, and it has been shown that the relative proportions of the major elements in solution are essentially constant throughout the oceans of the world. Sodium and chlorine, the two elements present in common salt, are by far the most abundant dissolved elements, and these, together with magnesium, sulfur, calcium, and potassium, account for 99.5 per cent of all dissolved solids in the sea (Fig. 2–4). Each cubic mile of sea water contains significant amounts of 64 other elements such as zinc, 47 tons; copper and tin, each 14 tons; silver, 1 ton; and gold, 40 pounds.

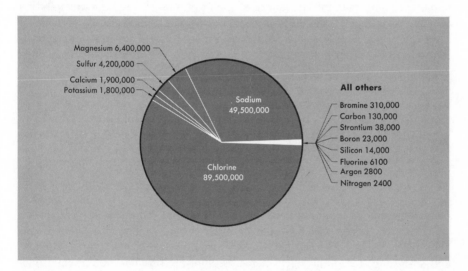

FIGURE 2–4   *Average amount of common elements in solution in a cubic mile of sea water. (Calculated by J. L. Mero, 1965.)*

Despite the wide range of elements present in sea water, only four are being commercially recovered in significant quantities at present: sodium and chlorine (recovered jointly in the form of sodium chloride), magnesium, and bromine. Sodium chloride (see Chapter 5) is locally produced by solar evaporation of sea water in shallow ponds. Bromine is recovered either by the addition of the complex organic compound aniline, which causes the precipitation of insoluble tribromoaniline, or by the addition of chlorine gas, which causes bromine to be released. Magnesium (see Chapter 3) is recovered by the addition

---

[1] For a discussion of the ocean composition, and for details of its configuration and chemistry, the interested reader should consult *Oceans* by K. K. Turekian, a companion volume to this one.

of dissolved calcium hydroxide causing the precipitation of magnesium hydroxide.

The composition of sea water with respect to its abundant elements is well established, and the volume of the ocean basins well understood. The sea's resources of the elements we now recover must therefore be classified as at least indicated reserves. The potential resources of the other elements, including those present in trace amounts, but which are not presently recovered, are also known with a high degree of certainty. Since the volume of the seas is about 330×10^6 cubic miles, the potential resources are truly enormous, and one must question why man has not exploited them more extensively. The reasons are numerous, and two difficulties are obvious. Unless specific reactions can be found to remove only the element or elements of interest, as in the precipitation of tribromoaniline, all the other dissolved compounds must also be removed— a wasteful process unless uses can be found for the large amounts of materials so produced. Furthermore, although the total amounts of minor elements present in the sea are large, the solution is still a very dilute one, and exceedingly large amounts of water must be processed to recover the small quantities dissolved. A point is soon reached where the cost of processing exceeds the value of the material produced. It has recently been estimated, for example, that even if extraction plants could process 1.2×10^6 gallons of sea water a minute, no element less abundant than boron could be profitably extracted because of the dilution levels. It is possible that new chemical methods might be found to extract some elements individually from the sea, but to most authorities the possibilities seem very remote. It is an interesting historical note that the German chemist and Nobel Prize winner, Fritz Haber, tried unsuccessfully for ten years to extract gold from sea water in order to repay German debts of World War I. We must conclude that for sodium, potassium, magnesium, calcium, and strontium; for the halogen elements chlorine, bromine, iodine, and fluorine; and for sulfur, boron, and phosphorus, the sea contains vast potential resources that may some day be exploited. For a large number of other elements the sea also contains vast amounts, but they are unlikely ever to be exploited because more profitable sources of the same elements can be found on land.

The Crust

The Earth's crust supplies most of our mineral resources. It has obvious and important differences from the hydrosphere and atmosphere. First, it is predominantly composed of *minerals* that are crystalline solids with specific and rather simple compositions. Second, as any walk through a rocky terrain will reveal, minerals are not randomly distributed, but are relatively concentrated in specific rocks and deposits. A limestone, for instance, contains mostly the mineral calcite, $CaCO_3$; a quartzite, mostly quartz, SiO_2; a lead vein, mostly galena, PbS; and a coal bed, mostly organic compounds. The chemical elements are therefore not evenly distributed through the crust, as they are in the atmosphere and oceans, but instead are distinctly segregated. In this fashion, even elements that have a low *average* concentration in the crust are sometimes found in exceedingly high *local* concentrations. The richest local concentrations are the ore deposits that man has exploited throughout his existence.

Owing to the difficulty of sampling beneath the deep oceans, the composition of the oceanic crust is not known with complete certainty. All indications are that it is everywhere basaltic and relatively constant in composition, and that the majority of the oceanic crustal rocks have never been elevated above the sea surface and subjected to erosion. There is no evidence to suggest that any sort of large-scale mineral concentrations do, or indeed should, occur in the oceanic crust; we must conclude that there is an exceedingly small probability of its containing mineral resources sufficiently important to warrant their recovery, though a possible exception to this statement may be found in certain mineral accumulations on the actual sea floor. An example of this is the extensive accumulations of nodular growth of manganese oxides on the floors of many of the world's ocean deeps. They are further discussed in Chapter 3.

The continental crust, which is somewhat more than half the mass of the whole crust, or about 0.29 per cent of the mass of the Earth, presents a sampling problem of a different kind. Owing to the wide range of rock types and their extreme geographic diversity, it is difficult to collect a reasonable number of samples so that accurate bulk compositional estimates can be arrived at by reliable averaging processes. Furthermore, only the surface of the continental crust is accessible, and this is largely covered by a thin veneer of *sedimentary rocks*, formed from the transported and deposited remains of weathering processes by waters of the hydrosphere. Geophysical measurements of the deeper portions of the crust, however, indicate a preponderant amount of more dense *igneous rocks* formed by the cooling and crystallization of magmas. Taking note of these uncertainties, K. K. Turekian has made a recent estimate of the continental crustal composition (Appendix I) and found that only nine elements account for 99 per cent of its mass (Fig. 2–5). If you compare Figs. 2–1 and 2–5, it is immediately apparent that such elements as silicon, aluminum, sodium, and potassium are relatively more abundant in the continental crust, whereas such elements as magnesium and iron are depleted relative to the rest of the Earth. It is not surprising, therefore, that rock types characteristic of the crust differ considerably from those of the mantle. Igneous rocks commonly occurring in the crust have characteristically high contents of silicon and aluminum, and are said to be *silicic*. Those occurring in the mantle, with much

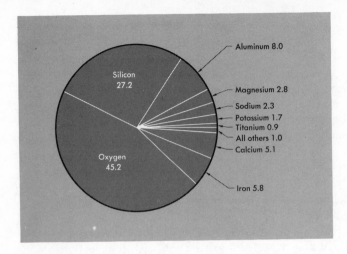

Silicon
27.2

Aluminum 8.0

Magnesium 2.8

Sodium 2.3

Potassium 1.7

Titanium 0.9

All others 1.0

Calcium 5.1

Oxygen
45.2

Iron 5.8

FIGURE 2-5 *Major elements in the continental crust. (After K. K. Turekian, 1969.)*

lower silicon and aluminum contents, but more magnesium and iron, are said to be *mafic.*[2]

There are still 79 elements to be accounted for, and as they total 1 per cent of the crust, they can only be present in trace amounts. Unfortunately, many of the elements that are vital resources fall in this category, as a glance at Appendix I will quickly show. It makes little sense to consider mining "average" rock, in which important resource elements have such low concentration, but fortunately there are a number of natural processes that have formed local concentrations, and from these man can draw—as he always has drawn—his resources.

[2] A full discussion of minerals and all rock types can be found in *Earth Materials* by W. G. Ernst, a companion volume to this one.

3

Metals:

the abundant elements

I have for many years been impressed with the steady depletion of our iron ore supply. It is staggering to learn that our once-supposed ample supply of rich ores can hardly outlast the generation now appearing, leaving only the leaner ores for the later years of the century. It is my judgment, as a practical man accustomed to dealing with those material factors on which our national prosperity is based, that it is time to take thought for the morrow. (Andrew Carnegie, Proceedings of a Conference of Governors in the White House, Washington, D.C., May 13–15, 1908.)

Metals have distinctive and versatile properties such as malleability, ductility, luster, and high thermal and electrical conductivities. They are invaluable for technological applications, and it has become standard for archaeologists and historians to use the metal-working skills of a community as one measure of its development; thus most of us are familiar with such terms as Bronze Age and Iron Age.

Metals can be readily divided into two classes on the basis of crustal abundance: the *scarce metals,* with abundances less than 0.01 per cent, and the *abundant metals*—iron, aluminum, manganese, magnesium, chromium, and titanium—with abundances greater than 0.01 per cent.

With rare exceptions such as a pure quartz sandstone, every rock contains detectable amounts of each abundant metal; these are usually present as separate minerals rich in the individual elements. It is the mineral form in which the element is carried that determines whether a given concentration is a useful resource. The abundant metals are most commonly combined with silicon and oxygen, the two most abundant elements of the crust, to form silicate minerals such as albite ($NaAlSi_3O_8$), anorthite ($CaAl_2Si_2O_8$), and olivine (Mg_2SiO_4). Silicate minerals are characteristically refractory and difficult to break down, however, so they are undesirable sources of their contained metals. The mineral forms in which the abundant metals are preferentially sought, therefore, are those from which the metals are most easily recovered; these are the oxides and hydroxides, such as *magnetite* (Fe_3O_4) and *goethite* ($HFeO_2$), commonly called *limonite* when very fine-grained and mixed with other compounds, and also the carbonates, such as *siderite* ($FeCO_3$) and *magnesite* ($MgCO_3$).

Consumption of the Abundant Metals

Because the abundant metals have high concentrations in the Earth's crust, they need not be greatly concentrated above their crustal averages before they can be economically mined by today's technological methods (see Fig. 4–2).

Consumption of the abundant metals is high, especially for iron and aluminum, and is continually increasing (Fig. 3–1). Because supply is plentiful and production large, the cost of the abundant metals is relatively low. For example, at the beginning of 1968 the price of pig iron in the United States was only 2.8 cents a pound, of aluminum 25.0 cents, and of magnesium 35.25 cents. Despite the ready availability of raw materials, production of the abundant metals requires advanced technology and large expenditures of power for major production. It is not surprising, therefore, that statistics (Fig. 3–1) show that the most technologically advanced country, the United States, is a very large consumer of each metal.

Iron

Iron, the second most abundant metal in the crust, is the backbone of modern civilization; it accounts for more than 95 per cent of all metals consumed. A significant proportion of the remainder—nickel, chromium, tungsten, vana-

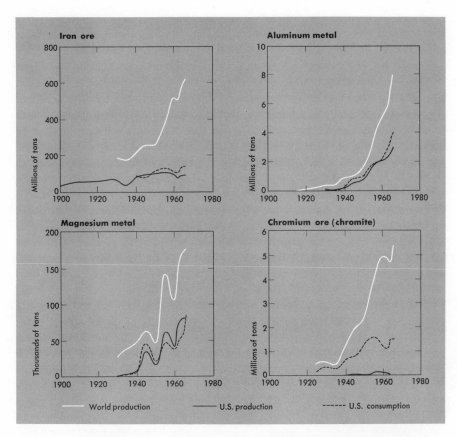

FIGURE 3-1 *World and U.S. production of several abundant metals, together with the U.S. consumption rate. The U.S., with six per cent of the world's population, consumes a much larger percentage of the total world production in each case, a pattern that holds true for most of the mineral resources presently produced. (After U.S. Bureau of Mines.)*

dium, cobalt, and manganese—are mined principally to be added to iron to give it more desirable properties of strength and resistance to corrosion. The ancient Egyptians knew and prized iron but probably did not discover the secret of its manufacture, relying instead on rare finds of meteoric iron. The first discoverer of the iron reduction process is unknown—there were possibly several discoverers in different parts of the world—but by 2000 B.C. the knowledge of ironworking seems to have become widespread, and by 800 B.C. most nations had already become skilled in its manufacture and handling.

The smelting of iron from its oxide ores is a chemically simple process in which carbon, in the form of coke, and iron oxide react at high temperature to form metallic iron and a gas, carbon dioxide. Because iron ore is rarely pure, limestone ($CaCO_3$) must also be added as a flux to combine with the impurities and remove them as slag. Selection of the best ore and production of sufficiently high temperatures are technological problems, and the history of the smelting of iron ores is the story of their solution. Until the beginning of the fourteenth century, all iron was produced in primitive forges by firing a charge of charcoal, iron ore, and a limestone flux in a blast of air. These forges, a form of which survive as the primitive Catalan forges, were small and capable of reducing the iron oxide to metallic iron but incapable of melting the reduced iron. They instead formed a pasty mass of incandescent iron grains which were welded together and from which slag and other impurities were removed by vigorous hammering to form wrought iron.

As the demand for iron increased, forges were made larger, and stronger air blasts were required to fire them. As a consequence, higher temperatures were reached and iron was produced in a molten form in furnaces that were forerunners of the modern blast furnace (Fig. 3–2A). By the fourteenth century,

FIGURE 3–2 *A modern blast furnace. (Left, A) Blast furnace J, rising 105 feet in the air at the Bethlehem Steel Corporation plant in Lackawanna, New York, produces 600,000 tons of pig iron annually. (Courtesy Bethlehem Steel Corporation.) (Facing page, B) The production of each ton of pig iron in a blast furnace from an ore containing 60 per cent Fe requires approximately 610 pounds of limestone as a flux and 2,440 pounds of coking coal. Recently developed electrical and oxygen furnaces require different mixes, particularly for being more efficient in their usage of coke, but the same three ingredients are needed.*

pig iron (as the raw furnace product is called) was being produced in large quantities, and the foundations of the modern industrial exploitation of iron were being laid. Subsequent discoveries led to: inexpensive ways to produce high-grade steels from pig iron; more efficient blast furnace procedures, including the use of coke made from coal to replace charcoal from dwindling supplies of wood; other methods to reduce iron ores; methods of handling ores containing deleterious impurities such as phosphorus and sulfur; and methods of processing low-grade ores as the richest were used up. Many of the more sophisticated developments occurred in the nineteenth century, laying the basis for today's technology in which almost all types of iron ores can be successfully handled, thus assuring the availability of truly enormous ore reserves to mankind. These developments have made transportation costs and access to markets a greater factor in the production of iron than for that of any other metal.

The production of each ton of pig iron requires approximately 610 pounds of limestone and 2,440 pounds of coking coal (Fig. 3–2B). Because the produc-

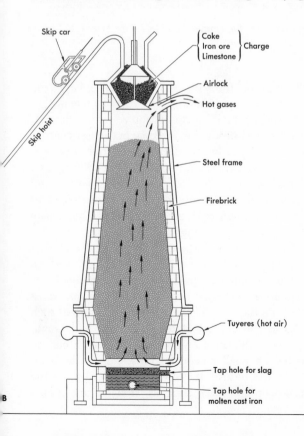

Skip car

Coke
Iron ore } Charge
Limestone

Airlock

Hot gases

Skip hoist

Steel frame

Firebrick

Tuyeres (hot air)

Tap hole for slag

Tap hole for
molten cast iron

B

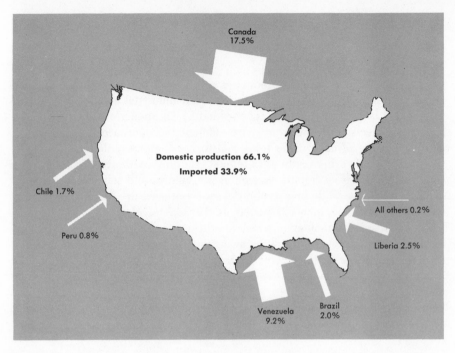

FIGURE 3-3   *Sources of iron ore consumed in the United States in 1966. (After U.S. Bureau of Mines.)*

tion of steel (and hence of the pig iron from which it is made) is large, the tonnage of additional materials is also large. Traditionally this has meant that the most profitable ores were those for which transportation costs were lowest. Britain's rise as the world's leading steel-producing country in the last century was due to the close proximity of high-grade coal deposits and rich iron ores. Construction of the St. Mary's River Canal at Sault Ste. Marie in 1855 opened the inexpensive water route of the Great Lakes and brought the largest and richest iron ores then known, those of the Mesabi Range in Minnesota, into easy accessibility to the rich Pennsylvanian coking coal deposits. Thus began the rise of the United States as the world's leading steel producer.

The transportation factor, though still important, is less critical now that large and efficient systems of transportation have been developed. Steel industries need no longer be set up as close as possible to the source of raw materials. Indeed, rich iron ores are now shipped by inexpensive water transportation from the far corners of the Earth to fill the growing demands of such major steel-producing countries as the United States (Fig. 3-3) and Japan, where production of rich ores is not sufficient to meet requirements. A natural consequence of this trend has been a smaller need for possession of the high-grade iron resources requisite to the industrial growth of a country, as the industrial development of iron-poor countries such as Holland and Japan attests.

Iron Resources

Iron is widespread in the Earth's crust and forms a great many minerals; of these, only four are important ore minerals (Table 3–1). The property of iron that most strongly affects its distribution is its ability to exist in more than one oxidation state. In the ferric, or trivalent, state it has a positive charge

Table 3–1

Important Ore Minerals of Iron*

Name	Composition	Content of Fe
Magnetite	Fe_3O_4	72.4%
Hematite	Fe_2O_3	70.0
Goethite (often called limonite)	$HFeO_2$	62.9
Siderite	$FeCO_3$	48.2

* Other iron-bearing minerals, such as pyrite, FeS_2, and chamosite, $Fe_2Al_2SiO_5(OH)_4$, occur in large quantities, often in direct association with the major ore minerals, but do not foreseeably constitute major resources of iron because of the technological difficulties of recovery.

of 3. In the ferrous state it is divalent, and in the metallic state, as in the Earth's core, it has a zero valence. The higher the positive valence, the more oxygen it combines with, because oxygen has a constant, negative valence of 2. Near the surface of the Earth, where oxygen is abundant, the ferric is the stable state; hematite and the dimorphous ferric hydroxides goethite and lepidocrocite are the stable minerals (dimorphous = same formula but different crystal structures).

There are three important classes of iron deposits:
1. *Deposits associated with igneous rocks*
2. *Residual deposits*
3. *Sedimentary deposits*

When a magma begins to crystallize after upward intrusion into cooler portions of the crust, several mechanisms may lead to deposits associated with the resulting igneous rock. A magma is a molten rock, a complex liquid containing many compounds; it does not crystallize at a fixed temperature as simple liquids do, but instead crystallizes over a temperature range, first one mineral forming, then another, until it is all solidified as an igneous rock containing several different minerals. If one of the early-formed minerals

Metals: the abundant elements

should be much more dense than the parent magma, it may sink rapidly and form a concentration by *magmatic segregation* on the floor of the magma chamber. Under rare circumstances the separating dense phase may be another liquid rather than a crystalline solid, and this new liquid, like the common example of oil and water, will separate into an immiscible zone or puddle. When both liquids have cooled and crystallized, the result is called a deposit due to *magmatic segregation by liquid immiscibility.* Early-formed minerals in a cooling magma tend to be anhydrous and free of the more volatile elements such as fluorine and chlorine. The residual magma becomes increasingly enriched in these volatile constituents which will eventually begin slowly to escape and possibly to alter, or *metamorphose,* the surrounding rocks. When immediately adjacent to intruding igneous mass, *contact metamorphic* deposits may be formed in this fashion. Finally, the escaping volatiles, which are commonly called *hydrothermal fluids* because they are hot and usually aqueous, may follow well-defined flow channels and, as they cool, deposit any dissolved matter they carry into *hydrothermal deposits.*

Iron deposits with igneous affiliations are usually contact metamorphic or magmatic segregation types. Contact metamorphic deposits are commonly small but very rich bodies of either hematite or magnetite. Few are large enough to be worked, though unusually large concentrations have been successfully mined at Cornwall, Pennsylvania (Fig. 3–4), Iron Springs, Utah, and Mount Magnitaya in the Ural Mountains of the U.S.S.R. The magmatic segregation deposits are typified by the great Kiruna deposit of northern Sweden. This great mass of magnetite, which outcrops for 1¾ miles in length over a width of 475 feet, lies more than 100 miles inside the Arctic Circle, in Swedish Lapland. Large-scale production began at the turn of the century and continues to the present day. The ore, which averages about 60 per cent

FIGURE 3–4 *Contact metamorphic deposits of magnetite near Cornwall, Pennsylvania, formed at the contact of a mafic igneous rock, called a diabase, intruded during the Triassic Period into sedimentary rocks of Cambrian age. (After A. C. Spencer, 1908, U.S. Geol. Surv. Bull. 359.)*

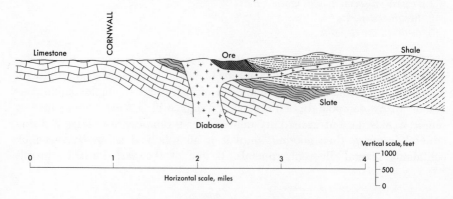

Fe, is largely magnetite, but also contains about 2 per cent P present in the mineral apatite, $Ca_5(PO_4)_3F$. It is assumed that the Kiruna ores were formed during the cooling of a gabbroic magma—a magma rich in iron, magnesium, and calcium, but relatively low in silica. During cooling of the gabbroic magma, an immiscible iron-rich liquid that contained a significant amount of phosphorus, but very little silica, separated out; being heavier than the gabbroic magma, it sank and accumulated. An apparently puzzling feature of the magnetite deposits formed by liquid immiscibility is that the temperature at which a pure magnetite liquid freezes is about $1,600°C$, far above the $1,050°-1,150°C$ temperature range where a gabbro freezes. The freezing temperature of the magnetite liquid, however, is depressed below that of the gabbro by the phosphorus, calcium, and fluorine of the apatite in solution, in the same way that salt depresses the freezing temperature of water.

Magmatic segregation bodies, like the contact metamorphic variety, are not widely distributed. Though the Kiruna ores are among the largest of all iron ore bodies and have been vital to the economy of Sweden, the scarcity of the class does not hold much promise of great undiscovered riches; thus they are not likely to be a major factor in the future of the world's iron resources.

Residual deposits of iron minerals are formed wherever weathering occurs and the ferrous iron present in a rock is oxidized to the relatively insoluble ferric form. This accounts for the brown, yellow, black, and red colors of weathered rocks and also for many of the soil colors we are familiar with. If the same weathering cycle removes more soluble minerals, the iron oxides and hydroxides remain concentrated as a residue. The process is known to have been active from Precambrian times to the present, and iron deposits formed in this way are very widespread in the geologic record. Commonly called brown ores, owing to the color of their main mineral constituent, goethite, the residual deposits were among the first to be exploited by man. Individual deposits of rich residual ores are small, however, often being only a few thousand or tens of thousands of tons, and do not lend themselves to the large-scale mechanized mining required by the modern iron industry. The importance of brown ores has therefore declined in recent times and will continue to do so.

It is worthwhile mentioning, however, that the iron-rich soils of the tropics, often too barren for concentrated agriculture because of the extreme tropical leaching, constitute a truly vast potential resource of iron. Though they are far too low-grade to be of use in meeting today's, or probably even tomorrow's, needs, these soils may eventually become our major source of iron.

The sedimentary iron deposits account for most of the world's production and identifiable resources. Though intensively studied for more than a hundred years, their origin is not yet completely understood. There are no known cases of the formation, today, of iron-rich sediments similar to those found in the geologic record; their origin must therefore be deduced from many different lines of evidence.

The sedimentary iron ores are nearly all *chemical sedimentary deposits;* that is, their constituents were transported in solution and deposited as chemical precipitates. Except for a few rare placer deposits of magnetite and hematite, they do not contain iron minerals carried as detrital grains. This fact presents the greatest dilemma in the puzzle. Weathering at the Earth's surface converts all the ferrous iron in a rock to the more stable ferric state, and the ferric iron compounds are essentially insoluble in most surface waters. The only way iron can be readily moved in the hydrosphere is to keep it in the more soluble ferrous state (which is almost impossible with the atmospheric reservoir of oxygen adjacent to it), or else somehow to change the normally neutral or slightly alkaline surface waters into acid waters, in which ferric iron is more soluble.

The period in the Earth's history when the greatest of the iron-rich sediments were laid down extended from 3.2 to 1.7 billion years ago. Called Lake Superior type ores after the area in North America where they were first studied, these ancient sediments are now found on all continents; because they were formed so long ago, however, we can only make educated geological guesses about how they formed. The conditions that best explain these important deposits include long periods of erosion and denudation of the continental masses and the occurrence of shallow inundations by the sea. The extensive erosion that preceded inundation left little detrital material to be deposited; thus, slowly-accumulating chemical precipitates were relatively more important in the new marine basins. Although we have no direct proof, it is believed that at this early stage in the Earth's history the atmosphere had a different composition. There was probably free oxygen present, yet many believe that the content of carbon dioxide may have been somewhat higher, perhaps even as much as 100 times higher, which still implies an abundance of about only 3 per cent. Under such conditions the surface waters would contain more carbon dioxide in solution and as a result would be slightly acid, allowing iron to be moved in solution and to be precipitated as iron oxide and hydroxide minerals in the shallow seas.

Transportation of iron in surface waters is only one of the interesting problems associated with the Lake Superior type iron ores. The sediments are laid down in fine bands—often in layers less than a millimeter thick—and are dull repetitions of iron-rich and silica-rich layers. So striking is this texture that the rocks are commonly called *banded iron formations* (Fig. 3–5). The silica-rich layers are now cherts—exceedingly fine-grained quartz resulting from the recrystallization of colloidal silica—and preserved within the cherts are the fossil remains of primitive microscopic plants that were apparently growing in the shallow seas of the time (Fig. 3–6). These microscopic plants,

FIGURE 3–5 *Photograph of a specimen of banded iron formation from the Mesabi Range, Minnesota. The chert-rich bands (light) which alternate with the iron-rich bands (dark) are believed by some to represent seasonal changes due to explosive growth of silica-secreting organisms during warm summer months.*

FIGURE 3–6 *Fragments of primitive microscopic plants approximately two billion years old, from the Gunflint Iron Formation, Ontario, magnified 1000 to 2000 times. (A) Spherical form with a regularly sculptured pattern. (B) Tubercle form in which tubercle-like spheres are attached to a larger sphere. (C) Filamental form with well-marked transverse septae. (Courtesy E. Barghoorn. From Barghoorn and Tyler, Science, v. 147. Copyright 1965 by The Amer. Assoc. for the Advancement of Science.)*

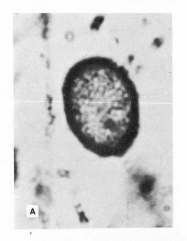

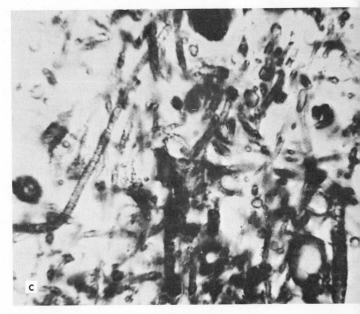

first found in the Gunflint Formation of Minnesota, but now identified in even the earliest banded iron formations around the world, are the oldest fossils known, predating the abundant fossil record beginning 600 million years ago, in Cambrian times, by at least 2,500 million years. Whatever the still unknown circumstances for the precipitation of banded iron formations, we are forced to conclude that they must not have been unusual, for the formations are found on all continents where rocks of suitable age occur (Fig. 3–7); thus we must further conclude that conditions suitable for their formation have not been present on Earth for at least 1.7 billion years, which is a significant fraction of the planet's history.

Although the great banded iron formations were formed in Precambrian times, many important iron-rich sediments are known in the post-Cambrian epochs, a 600-million-year period in which the continuity and evolution of animal and plant life provides good evidence that the atmosphere remained nearly constant and similar to that of today. The post-Cambrian deposits have important differences from the Precambrian, however. They are usually associated with abundant detrital material, although the iron minerals themselves are clearly chemical precipitates and are limited in areal extent. The Precambrian iron sediments can be traced for hundreds and sometimes thousands of miles and apparently formed over a large part of the shallow sea floor. The post-Cambrian deposits formed only in restricted basins, often less than 50 miles across, and according to H. L. James show clear evidence of being due to these special and local, rather than worldwide, conditions: First, a warm and humid climate which allowed a thick cover of deeply rooted plants to develop on the land surface, and second, soils and permeable rocks penetrated by ground waters rich in carbon dioxide from the subsurface production by root processes, and rich in organic acids from plant decay. The resulting subsurface waters were quite acid and leached iron from the soil and rocks. The leached iron moved out via subsurface routes to the sea. When the debouching area was the open ocean, the iron was quickly dispersed, but in special cases where the debouching area was a shallow, restricted marine basin, the iron was trapped and accumulated as a chemical sediment.

Iron ores of the post-Cambrian type have been very important in the past, supplying most of the iron mined in Great Britain, France, Germany, and Belgium, as well as an important fraction of the Newfoundland and Birmingham (Alabama) ores. Their worldwide importance is decreasing, however, as the European deposits are depleted and an ever greater reliance is placed on the Lake Superior type ores.

The Precambrian banded iron formations contain from 15 to 40 per cent Fe, traditionally considered too low to warrant recovery. However, where the iron formations have been elevated and exposed at the surface, chemical weathering has often removed the associated siliceous or carbonate minerals

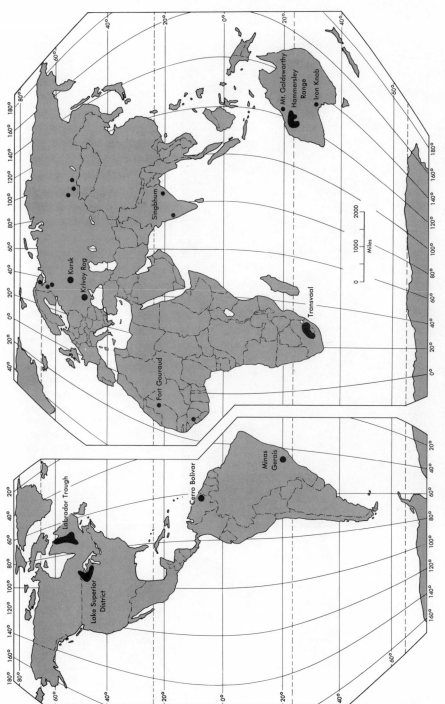

FIGURE 3–7 Location of the major Precambrian banded iron formations of the Lake Superior type. These ores assure abundant iron resources for centuries ahead.

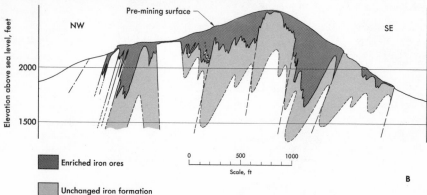

FIGURE 3–8 (A) Aerial view of Cerro Bolivar, Venezuela, and the mining operations on its summit. The hard and resistant mass of banded iron formation stands above the surrounding plain as an erosion remnant. (Courtesy Orinoco Mining Company.) (B) A cross section through the hill of Cerro Bolivar, showing the rich iron ores formed by secondary enrichment of the original iron formation. (After J. C. Ruckmick, 1963, Economic Geology, v. 58, p. 222, with permission of the publisher.)

and left a secondarily enriched residual ore containing 55 per cent Fe or more (Figs. 3–8A, B). The great Precambrian iron deposits of the Lake Superior region and the Labrador Trough in North America, of Cerro Bolivar in Venezuela and Minas Gerais in Brazil, of Krivoi Rog in the Ukraine, and in many other parts of the world are all extensive iron formations with local zones enriched by the leaching of silica. Until recent years the unleached and therefore unconcentrated iron formations below the rich ores had not been mined, but with the depletion of the concentrated ore in the United States, efficient ways have been developed to mine and beneficiate the unleached ores—called *taconites* in the Lake Superior district. *Beneficiate* is commonly used in the mining industry and means to free a mineral from its enclosing waste rock and to effect an inexpensive concentration.

The mining of low-grade taconites is a good example of the need to consider technological advances in estimating potential resources. When first considered after World War II, taconites were thought of as expensive—even as desperate—alternatives to dwindling supplies of the rich ores in the Lake Superior district. When production and beneficiation of the lean ores were begun, however, it was quickly found that in many instances the concentrated pellets that were produced made a better blast furnace feed than the traditional ores, and that consequent savings in smelting more than offset the extra costs of mining and beneficiation. The taconite pellets have now become the standard of quality in the industry; by 1968 they accounted for 40 per cent of the iron produced in the United States. The trend will certainly continue, and it has been estimated that by 1978 it will account for more than 75 per cent of production.

The known reserves of iron ore available in the leached and enriched ores are large—many billions of tons—but are miniscule compared to the amount present in the unaltered iron formations. Estimates by the U.S. Geological Survey in 1965 showed that taconite reserves in the Lake Superior region alone exceeded 10^{11} tons of Fe. Other estimates revealed that even larger potential resources—in excess of 10^{12} tons at each place—were available in the iron formations of the Transvaal, the Ukraine, the Hammersley Ranges of Western Australia, the Labrador Trough of northeastern Canada, and Minas Gerais in Brazil.

The world's potential resources of iron are so greatly in excess of present and projected needs, assuming the most liberal growth rates of production, that it seems safe to assert that many centuries will pass before depletion of the ores becomes a serious problem. Mr. Carnegie's pessimistic remarks quoted at the beginning of the chapter have been proved unwarranted by man's technological advances.

Aluminum

Aluminum is even more abundant in the Earth's crust than iron and, because of its lightness, has a more desirable weight/strength ratio than iron. Although the element was first separated in a pure form in 1827, it was not until the end of the nineteenth century and the early part of the twentieth that methods capable of producing aluminum metal of high purity were developed. From that time on, the production and number of uses of aluminum have steadily increased to the point where the world's production exceeded 8 million tons in 1968. Aluminum has, besides its desirable strength and weight properties, a high resistance to corrosion; in addition, it is a good conductor of electricity. Most of the technological uses take advantage of one or more of these properties, and, as a result, aluminum has challenged other metals, such as iron for some of its structural purposes and copper for some of its electrical uses; it is now estimated, for example, that more than 90 per cent of the new electrical transmission lines in the United States contain aluminum conductors. This element has its own distinctive properties, however, and has proved highly versatile in new uses in the construction and transportation industries, its two largest consumers.

The production of aluminum requires exceptionally large expenditures of electrical power; in 1965 the aluminum industry consumed approximately 3 per cent of the power generated in the United States. As production grows, it becomes increasingly desirable to ship aluminum ores to sources of inexpensive electrical power, such as the hydroelectric sources along the Columbia River in the Pacific Northwest and the huge sources in northern Canada. This high-power usage has been the most important feature in determining the distribution of aluminum producers around the world.

Aluminum Resources

With the exception of the rare mineral nepheline ($NaAlSiO_4$), mined in the northern Soviet Union, the only aluminum minerals exploited on a large scale to date are the hydroxides (Table 3–2). The world reserves of aluminum in the form of hydroxide-rich ores are limited, however, and very erratically distributed. It is already apparent that other common aluminum minerals will have to be used for future production, despite the technological difficulties involved. The possible candidates for alternative supply sources are confined to the minerals listed in Table 3-2.

Minerals formed by igneous processes and by metamorphism are stable

Table 3-2

Currently and Potentially Important Ore Minerals of Alumium*

Mineral	Composition	Content of Al
Boehmite, diaspore	$HAlO_2$ (dimorphs)	45.0%
Gibbsite	H_3AlO_3	34.6
Andalusite, kyanite, sillimanite	Al_2SiO_5 (trimorphs)	33.5
Kaolinite (the most aluminous clay)	$Al_2Si_2O_5(OH)_4$	20.9
Anorthite (the most aluminous feldspar)	$CaAl_2Si_2O_8$	19.4
Nepheline	$NaAlSiO_4$	18.4

*Discussion of potential resources of Aluminum may be found in Bulletin #1228 of the U.S. Geological Survey, 1967.

deep in the Earth's crust, an environment where large bodies of free water do not exist; such minerals are often anhydrous, or at best contain very little water. When brought to the surface, they are no longer stable; although the rate of chemical change to a more stable form is slow, they are gradually transformed into new minerals, most of which are hydrous. This transformation at the Earth's surface is called *chemical weathering*. During chemical weathering, elements such as Na, K, Ca, and Mg form relatively soluble compounds and are soon removed. The residue, left as a capping, is called a *laterite* (Fig. 3–9).

FIGURE 3–9 *Leaching of the most soluble components during chemical weathering leaves a lateritic capping in which the less soluble aluminum and iron hydroxides are concentrated. When aluminum hydroxides predominate, the laterite is called a bauxite.*

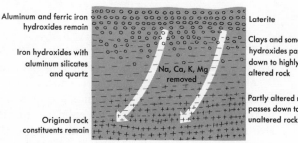

Aluminum and ferric iron hydroxides remain — Laterite

Iron hydroxides with aluminum silicates and quartz

Na, Ca, K, Mg removed

Original rock constituents remain

Clays and some hydroxides pass down to highly altered rock

Partly altered rock passes down to unaltered rock

Most laterites are iron-rich, but some are aluminum-rich and are called bauxites (after the little village of Les Baux in southern France, where they were first recognized in 1821).

Bauxites apparently form in only one way—as residual deposits due to weathering under tropical conditions. On a low-lying or relatively flat surface, rainwater runoff is slow, and mechanical removal of weathering products in suspension is minimal. The only way material can be moved under such circumstances is in solution, and the solubility of most minerals is extremely low. However, in high rainfall areas the volume of water available for solution is large; in the warm climates found in tropical zones, which speeds up the rate of solution, vast amounts of material can eventually be removed. Besides rainfall, the topography, temperature, and acidity of the leaching waters are believed to be important in the formation of bauxites. After the removal of most elements, a residue rich in both silica and aluminum hydroxide remains, usually in a combined form such as the clay mineral kaolinite $(Al_2Si_2O_5(OH)_4)$. Kaolinite, too, will dissolve. If the water percolating down is very acid, it goes into solution completely; if the water is very slightly acid, only the silica component of the kaolinite goes into solution, and the alumina is left behind as bauxite.

The source rocks for bauxites have a wide range of compositions, some of which do not contain much aluminum to begin with, but most tend to have relatively low silica contents compared with other rocks. In fact, at least half of the world's bauxites are developed from limestones, which are essentially calcium carbonate, but which usually contain small amounts of clays and even smaller amounts of iron minerals, so that the laterites formed from them are bauxitic (Fig. 3–10). In a tropical climate, limestone dissolves rapidly and the

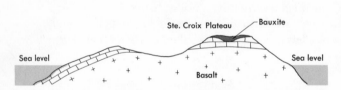

FIGURE 3–10 *A residual bauxitic capping left after long-continued leaching of limestone in the Ste. Croix District of Haiti. (Modified after O. C. Schmedeman, 1948.)*

clay residues remain; the acidity of the waters leaching the limestone is apparently just right for the development of bauxite from the clays.

A rock type on which very rich bauxites developed in an area near Little Rock, Arkansas, is nepheline syenite, a coarse-grained igneous rock composed largely of feldspar minerals and nepheline, but devoid of free quartz. Extensive leaching in a tropical climate during the Eocene Epoch developed the bauxites that are mined today. A classic study of the Arkansas deposits by W. Mead in 1915 first revealed the essential features of the leaching process. Nepheline broke down to form clays such as kaolinite, then the feldspars followed similarly. When mostly clays remained, these in turn began to break down to gibbsite, the primary mineral of Arkansas bauxites. Of course, the leaching process removes much material in solution too, so the rock becomes increasingly porous; in the end, no more than half the mass of the original rock remains (Fig. 3–11).

Bauxites are widespread in the world but are concentrated in the tropics. Even where they are found in presently temperate conditions such as southern France, it is clear that at whatever time in the past they did form, the climate was more tropical. Because they are superficial deposits, they are exposed and vulnerable to mechanical weathering processes if climatic or geologic con-

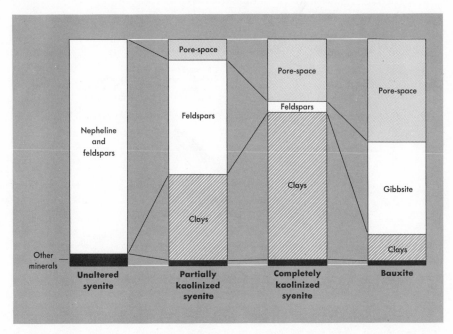

FIGURE 3–11   *Successive stages in the leaching of nepheline syenite to give the Arkansas bauxite ores. (Modified from W. Mead, 1915, Economic Geology, v. 10, p. 48.)*

ditions should change in any way. Bauxites are very rare in glaciated regions, for example, because the overriding glaciers scrape all the soft materials off the surface. Because of their vulnerability to later erosive processes, most of the known bauxite deposits are geologically young: more than 90 per cent of them formed no earlier than the Cretaceous Period, and the largest of all formed in the modern tropical regions during the last 25 million years.

Tropical regions are among the least developed parts of the world, and it is only in the period following World War II that they have been extensively explored for bauxites. The results have been spectacularly successful, with the discovery of vast reserves of rich ores in the tropical regions of Australia, the

Caribbean, South America, and Africa. The world's reserves, as estimated by the U.S. Geological Survey in 1967, are given in Table 3-3. The total of all kinds of bauxite reserves is 16.4 x 10^9 tons. Although this is very large, it is far less than the huge potential resources of iron. Even if discovery is made of large new supplies of bauxite in tropical regions, the potential resources are unlikely to challenge those of iron. While bauxite reserves are more than

Table 3-3

Known World Reserves of Bauxite

Region	Measured and Indicated Reserves; Tons × 10^6	Inferred Reserves; Tons × 10^6
North America	45	300
Central America	—	200
Caribbean (mainly Jamaica)	690	450
South America (mainly Surinam and Guyana)	390	1000
Europe	850	400
Africa (mainly Guinea and Cameroon)	1580	4800
Asia (mainly China)	280	1400
Oceania (mainly Australia)	2000	1000
Total	5800	9600

After S. Patterson, U.S. Geological Survey, Bulletin #1228, 1967.

adequate for the immediate future—even for some time beyond the end of the present century—they are clearly incapable of sustaining the aluminum industry far into the future. Furthermore, the inaccessibility of many of the present known deposits, in addition to the difficulty of working in the tropics, has already ensured the development of alternative sources of aluminum. Of these, the most likely is to be found in clay minerals of the kaolin group, which form extensive sedimentary and residual deposits.

Clays are formed during the weathering cycle of most rocks containing aluminum and, following quartz, are the most abundant minerals in newly deposited sediments. Once the technological difficulties of aluminum recovery from clays are overcome, the world's resources of aluminum will be almost limitless. This day may not be far off, because pilot plants for the production of aluminum from relatively pure clay beds are already operating in the United States and can produce aluminum that is only 25 per cent more expensive than that produced from bauxite.

Manganese

Manganese is essential for the removal of oxygen and sulfur in the production of steels. Fourteen pounds of it are required for every ton of carbon steel produced, and no satisfactory substitute has ever been found.

Like iron, manganese can exist in more than one oxidation state, and its distribution is controlled by this property. Manganese is readily concentrated by sedimentary processes; in fact, all the important manganese resources of the world are in sedimentary rocks or residual deposits formed by leaching them. As with iron, the most oxidized compounds of manganese are the least soluble and are those concentrated in residual deposits.

The world's greatest known deposits of manganese, in which pyrolusite (MnO_2) occurs in chemical sedimentary deposits, are found in the Soviet Union, principally at Chiaturi in Georgia and at Nikopol in the Ukraine. These two deposits contain hundreds of millions of tons of measured reserves. Important residual deposits, mainly of pyrolusite and psilomelane $(Mn_2O_3 \cdot 2H_2O)$, are found in India, Ghana, Brazil, Egypt, and the United States. Like bauxite, the rich residual deposits of manganese oxides are concentrated in the tropics, where high rainfall and deep weathering are common. Though widely distributed, the world's inferred reserves of manganese ores were reported by the U.S. Bureau of Mines to be only about 2×10^9 tons in 1965. Compared with a present total annual production of less than 20×10^6 tons, however, this is a large figure; it also reflects considerable optimism, since it is highly probable that further exploration in tropical regions will uncover large sources of rich manganese ores.

One of the largest potential resources of manganese, and one that has excited much discussion and speculation in recent years, rests on the ocean floors. During the years 1873–1876 the Royal Society of London sent the ship "Challenger" on an epoch-making voyage around the world to gather data on the waters, animals, plants, and bottom deposits of the oceans. One of the most interesting discoveries made was of an abundance of black manganese oxide nodules, as much as several inches in diameter, covering the floors of the three major oceans (Fig. 3–12). Although these nodular deposits have been intensively studied in recent years, their origin remains something of an enigma. The manganese is apparently derived by normal erosive processes on the land surface and then, through a complex series of steps, slowly migrates to the ocean bottom, where it accumulates. The rate at which the nodules grow appears to be extremely slow—no more than 0.01 mm per 1,000 years. They are widely spread on the sea floor, however, and it has been estimated that ap-

FIGURE 3-12　*Manganese nodules on the sea floor at N 31° 20', W 67° 36' in 5026 meters of water. The nodules are from 1 to 5 cm in diameter and are found in the deep oceans around the world. (Courtesy Bruce Heezen, Lamont Geological Observatory.)*

proximately 1.7×10^{12} tons of nodules could be dredged from the bottom of the Pacific Ocean alone. Nevertheless, the technological problems of recovery are so vast, and the probable disruption of natural processes in the ocean is so great, that most scientists and engineers who have considered the problem conclude that the exploitation date is too far in the future to warrant present speculation.

Chromium

Chromium, like manganese, is an essential alloying metal used in the steel industry, and one for which satisfactory substitutes have never been found. In recent years more than 60 per cent of all chromium production has been used by the metallurgical industry. Important amounts are also consumed in the chemical industry and in the use of chromite, $(Mg,Fe)_2CrO_4$, the only ore mineral of chromium, for refractory bricks.

Until the end of the last century, chromium was used only in the chemical industry, principally for pigments. After 1900, however, it gained importance as an alloying element; it is steadily growing in use as high-speed tool steels and stainless steels find wider application.

All known chromite deposits are associated with mafic igneous rocks, such as dunite, which is very rich in olivine, and peridotite, a rock containing pyroxene and feldspar minerals in addition to olivine. Chromite deposits apparently form only as magmatic segregations, since chromite, being very dense, sinks to the bottom of the magma chamber where it accumulates in rich pods and layers. Although chromite deposits are found in all parts of the world, more than 80 per cent of the world's production of 5,110,000 tons in 1967 came from only five countries—Republic of South Africa, U.S.S.R., Turkey, Southern Rhodesia, and the Philippines (in decreasing order of importance). Inferred reserves of rich chromite are large, especially in South Africa, where one of the most unusual masses of igneous rocks in the world occurs. Known as the Bushveldt Complex, and comprised of many different layers of mafic igneous rocks several of which contain chromite, this mass covers hundreds of square miles and contains reserves estimated in the hundreds of billions of tons. It is anticipated that South Africa will increasingly become important, and eventually dominant, in the world's production of chromite.

Titanium

The metal titanium, like aluminum, combines light weight with high strength and high resistance to corrosion. For some purposes—for example, in the space industry and in the development of supersonic transport planes—it has answered many technological needs. It is a difficult metal to work, however, and an even more difficult metal to recover from its ores. Indeed, commercial production of titanium began only after World War II, and despite a yearly increase of 10 to 15 per cent, U.S. consumption of it had only reached 20,100 tons by 1967. Although it has been confidently predicted that this growth rate will continue, it is still too early in the history of titanium to estimate accurately its ultimate importance to mankind.

By far the largest use of titanium, accounting for more than 90 per cent of world consumption, is as the oxide (TiO_2), which is widely used as a white pigment, particularly for paints.

Ilmenite ($FeTiO_3$), the main ore mineral, is apparently concentrated by magmatic segregation from a cooling anorthosite, an igneous rock rich in anorthite ($CaAl_2Si_2O_8$) and often associated with gabbros. The largest known titanium deposit in the world, at Allard Lake, Quebec, is of this kind, as are the deposits at Sanford Lake in the Adirondack Mountains of upper New York State and at Blaafjeldite in Norway.

Both ilmenite and rutile (TiO_2) are widely distributed in small amounts in igneous and metamorphic rocks; both are heavy and extremely resistant to chemical and mechanical erosion. This resistance to breakdown puts them among the last minerals to disappear in the erosion cycle; they thus tend to concentrate, with other heavy resistant minerals, in *placers* (see Chapter 4). In Kerala, on the southwest coast of India, there are 15 miles of ilmenite-rich beach sand placers, and these are beneficiated by simply removing the ilmenite magnetically. Other important placer deposits are known along the southeast coastline of the United States, the coast of central Brazil north of Rio de Janeiro, the coasts of northern New South Wales and southern Queensland in Australia, and off Honshu, in Japan.

The Canadian deposits, which reportedly contain more than 50 million tons of titanium, and the enormous beach sand deposits, such as those at Honshu (with an estimated 10 billion tons of sands carrying about 7 per cent titanium), appear more than adequate to meet man's demands even into the distant future.

Magnesium

Magnesium is the lightest metal and, being strong, is in demand for production of light corrosion-resistant alloys. The annual production of magnesium as a metal, however, is small compared with that of iron and aluminum—about 203,000 tons for the world in 1967. The main use of magnesium is in compounds, particularly the oxide (MgO), which has desirable thermal and electrical insulating properties. The major sources of magnesium are the sea, which contains an inexhaustible supply (see Chapter 2), and the minerals dolomite, $CaMg(CO_3)_2$, and magnesite, $MgCO_3$, both of which are widespread constituents of the crust. Dolomite occurs in marine sedimentary rocks, and commonly forms an essential constituent of a rock called dolostone. Magnesite is found in sediments, residual concentrations, and hydrothermal deposits.

Reserves of magnesium are almost limitless and are so widely available to all nations that any discussion of their magnitude has little point. The only limits and constraints on man's use of magnesium will be his own technology.

4

Metals:

the scarce elements

I believe that the prospect of impending shortages . . . will continue to inspire the . . . technical advances that will make it possible to resolve . . . the doom we often are prone to foresee. (T. B. Nolan, The Inexhaustible Resource of Technology, *1958.)*

We have defined the geochemically scarce metals as those with crustal abundances below 0.01 per cent. It is surprising to find that such common commodities as copper, lead, zinc, and nickel, all of which have large and growing rates of production (Fig. 4–1), are geochemically rare and belong in the scarce category with gold, silver, and platinum. Most experts believe that it is in this group of resources that shortages are likely to develop first, posing a serious challenge to technological development. The scarce metals are the group of vital resources that speeded development of the technological marvels of the last hundred years, such as the generation and distribution of electrical power; telegraphic, radio, and television communication; aeronautics; rocketry; and nuclear power. Development of the electrical industry, for example, would certainly have been much slower without abundant and inexpensive supplies of copper.

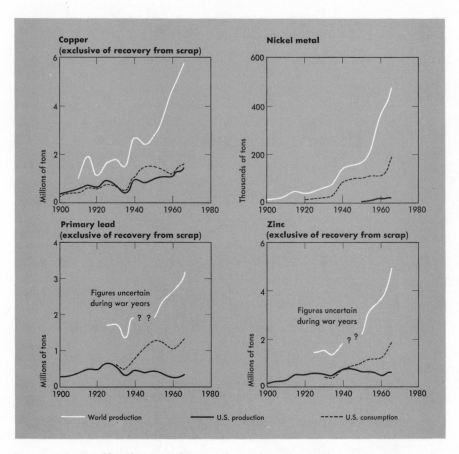

FIGURE 4-1 *World and U.S. production of several geochemically scarce metals, together with the U.S. consumption rate. The highly industrialized U.S. consumes at least 25 per cent of the world's production of each metal. (After U.S. Bureau of Mines.)*

The relative abundances of scarce metals are low, but the total amounts in the crust are large simply because the crust itself is large. Although some have endorsed the idea, it does not make sense to consider "average" rock as a potential source of scarce metals; the huge tonnages of rock to be processed and the tremendous power consumption needed to conduct the operation are prohibitive of profit. Man has always sought *ore deposits,* the localized geological circumstances in which scarce metals are highly concentrated and from which they can be won cheaply and rapidly. A number of factors determine the amount of concentration needed for an element to be profitably recovered. With all factors favorable, the present minimum concentrations are high (Fig. 4-2), though, as we shall see later, the minimums have been steadily reduced

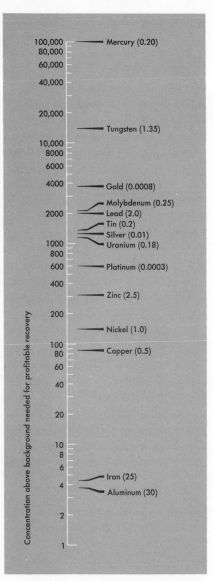

FIGURE 4-2 *The crustal abundances of the geochemically scarce metals are so low that very large concentrations above background are needed before deposits can be profitably mined. The abundant metals require much lower concentration factors to produce rich ores. The bracketed percentages are the minimum metal contents that an ore must have before it can be mined under the most favorable circumstances with present-day technology.*

for some metals, such as copper, by new and efficient mining (see Fig. 4-5). We do not yet know how low the necessary concentration factors can be pushed before the cost of recovering the metals is so high that cheaper substitutes such as cements, ceramics, plastics, or even abundant metals will take their place. Technology too, which relies on scarce metals for so many special needs, is a resource—of man's own ingenuity— and can be vitally important in evaluating future mineral resources. What is considered hopelessly impractical for exploitation by today's standards, or even by those projected for tomorrow, may well be tapped in the future, owing to suitable technological advances. It is sensible to be optimistic, like Dr. Nolan, and believe that a successful technology will learn to supply its own needs and will continue to grow.

Geological Distribution of the Scarce Metals

The scarce metals are widely distributed, but unlike the abundant metals they rarely form separate minerals. Instead, they reside in the structures of common rock-forming minerals, usually the silicates, an atom of a scarce metal substituting for an atom of an abundant element. For example, nickel atoms substitute for those of magnesium in olivine, $(Mg_2SiO_4$—Fig. 4-3), though commonly only to the extent of a few nickel atoms for every million atoms of magnesium. Substitution of foreign atoms causes strains in a mineral structure, and accordingly there are limits to the process; these are determined by temperature, pressure, and various chemical parameters related to the rock composition. For most common rocks and minerals the limits are not exceeded, and the scarce metals remain atomically locked in the host structures. To recover them, the host mineral itself must be broken down, and this is an expensive chemical process because, as we have seen, silicate minerals are highly refractory and difficult to reduce. When the limits are exceeded, however, the substituting element forms a separate mineral—for example, pentlandite, $(Ni,Fe)_9S_8$, in the common case of nickel; the way is then open for an inexpensive beneficiation process. The physical properties of pentlandite differ markedly from the associated silicate minerals, and simple crushing, followed by a concentration based—for example—on specific gravity or surface property (flotation) differences, will produce an inexpensive, nickel-rich concentrate of pentlandite before an expensive chemical reduction is needed. The

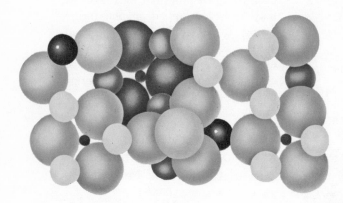

FIGURE 4–3 *Atomic structure of olivine (Mg_2SiO_4), a mineral found in many mafic rocks and in which small amounts of nickel may be carried by atomic substitution. The large spheres are O atoms, the smallest Si. Intermediate-sized spheres are Mg atoms, and these are randomly replaced by Ni atoms (black) of essentially the same size. (After W. L. Bragg, 1937.)*

principle of having a scarce metal present in a separate mineral, which in turn has distinctly different physical properties from its associated minerals, has always been the most important factor in the utilization of the metals and the exploitation of their ores. It is significant, for example, that the scarce metal gallium, with a crustal abundance approximately twice that of lead, occurs almost exclusively as a substituting element and has never been of vital importance in technology. Had inexpensive and obvious sources of easily reduced gallium minerals been available, man would undoubtedly have found a myriad of uses for them. It is unfortunate that only a tiny fraction of the scarce metals in the crust—less than 1 per cent—occurs other than as structural substitutes in common minerals. The substitution principle has one advantage, however, as exemplified by the metal indium. The world's consumption of indium is only 20 tons per annum, but this metal forms only one known rare mineral of its own, and no individual mine or deposit has ever been found. Indium commonly substitutes for zinc in sphalerite (ZnS), the common ore mineral of zinc, and can therefore be recovered as a bonus or by-product in the smelting of sphalerite.

Other elements such as silver and cadmium, which, because of their low abundance levels, rarely form separate minerals, are also produced largely from ores where they are carried by atomic substitution. Silver, for example, commonly substitutes for copper in the ore minerals tetrahedrite ($Cu_{12}Sb_4S_{13}$) and chalcocite (Cu_2S), and for lead in the mineral galena (PbS). Recovery of silver as a by-product is now so large that in 1966, of the ten largest silver producers in the United States, five were primarily lead and zinc producers, and three primarily copper producers, whereas only two—both in Idaho—mined their ores primarily for the silver content.

Metallogenic Provinces

Deposits of scarce metals are formed in many ways and may, at first glance, seem to be haphazardly distributed. This is not so, for closer study shows marked tendencies for deposits of one or more metals to be grouped in certain geographic belts. These belts, with their high rate of ore deposit occurrence, extending for perhaps hundreds or even thousands of miles, have been called *metallogenic provinces* (Fig. 4–4) and in many cases remain geological mysteries. One puzzle, for example, is that no known evidence exists to suggest that the rocks in a given province are actually richer in scarce metals than an equivalent, but deposit-barren, volume of rocks elsewhere in the Earth. Rather, it appears that processes forming deposits occur more frequently in the provinces than outside them, but that the average composition of the crust is essentially the same everywhere. To understand why provinces form, why individual deposits are where they are, and in order to seek for other, presently concealed, deposits, we must learn the whole geologic history of the province and isolate the ore-forming factor or factors. These goals remain as vital, fascinating, and only partly solved problems for geologists.

Discovery of New Ore Deposits

The tremendous consumption of scarce metals in the last two centuries resulted from the most incredible expansion in man's history, a period that cannot be taken as the standard for the future. As Western man expanded into, and explored, the Americas, Africa, Asia, and Australia, he found abundant ore deposits. Unfortunately, there is no evidence that discovery rates of such rich, easily found deposits will long continue; on the contrary, in the more

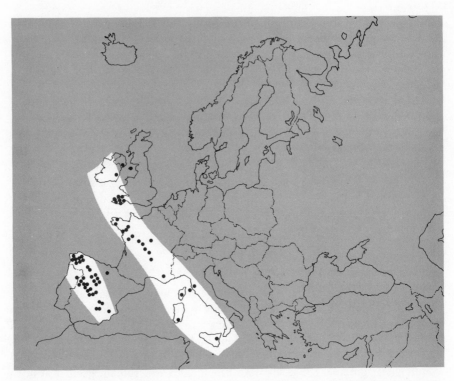

FIGURE 4-4 *An example of two metallogenic provinces in Southern Europe, where deposits of tin minerals (dots) are found in two well-marked belts. (After R. D. Schuiling, 1967, Economic Geology, v. 62, with permission of the publisher.)*

carefully prospected areas such as the United States and Europe, the discovery rate has already decreased drastically. It was recently pointed out by R. J. Forbes, for example, that the mines operating in the portion of Europe formerly embraced by the Roman Empire had all been known to the Romans. The only exceptions were deposits of metals such as aluminum and chromium, metals that the Romans did not use. Despite close settlement and keen observation by members of a society aware of the importance of metals, no major discoveries were made during a period of almost 2,000 years.

As the inadequately prospected areas of the world diminish, we are forced to develop more sensitive means of searching beneath soil and overlying rock covers and to develop criteria for narrowing the areas of search to those few selected spots on Earth where the probability of finding ores is greatest. These two goals now occupy the attention of an increasingly large body of able scientists. Their success or failure will directly determine the extent of man's future use of the scarce metals.

Classification of Scarce Metals

When scarce metals form ore deposits, the minerals in them have distinctive properties and compositions, and on this basis they can be grouped in three categories. The first, which includes Cu, Pb, and Zn, commonly forms *sulfide minerals*. The second, which includes Au and Pt, commonly occurs as *native metals*. The third, which includes W, Ta, Sn, Be, and U, commonly forms *oxide* and *silicate minerals*. Some overlap occurs—Sn, for example, forms both sulfide and oxide minerals—but the groupings take note of the major and most important minerals.

Scarce Metals Commonly Forming Sulfide Deposits

The number of scarce metals concentrated principally in sulfide deposits is large—Cu, Pb, Zn, Ni, Mo, Ag, As, Sb, Bi, Cd, Co, and Hg, together with numerous rarer ones that occur largely or solely as atomic substitutes for other scarce metals. We shall discuss only the most important of these.

Copper

Copper, a metal used since antiquity, but now the workhorse of the electrical industry because of its excellent properties of conduction, is almost common enough to be classed as an abundant metal. Copper deposits are widespread, but mostly as veins or contact metamorphic deposits of a million tons or less, which is small for purposes of mining; they are often exceedingly rich, however, and many exceed 10 per cent copper. Until the turn of the present century, such deposits accounted for most of the copper produced in the world, and because of the high cost of working small deposits by underground mines, the lowest-grade ores that could profitably be mined were about 3 per cent. Although copper is still mined on all continents, production from vein deposits has continuously declined in importance as the discovery and exploitation of large, low-mining-cost deposits, such as the *porphyry copper deposits,* have proceeded. These large-volume deposits have permitted the workable grade of copper ore to be steadily diminished to today's low level of 0.5 per cent (Fig. 4–5).

Porphyry coppers are large, low-grade hydrothermal deposits containing at least 5 millions of tons of ore averaging 2 per cent Cu or less, in which the copper mineral, usually chalcopyrite ($CuFeS_2$), is so evenly distributed that large-volume, and consequently inexpensive, mining practices can be em-

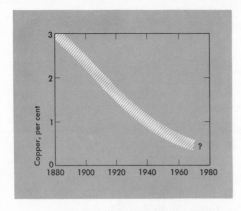

FIGURE 4–5 *A steady reduction in the minimum grade of copper ores that can be profitably worked has arisen from the discovery of large, low mining-cost deposits. (After U.S. Bureau of Mines.)*

ployed. The usual system employed is surficial mining from huge open pits (Fig. 4–6). These deposits are typically associated with an igneous intrusion of variable, but always silicic, composition. The intrusions have a characteristically porphyritic texture of large feldspar or quartz crystals set in a much finer-grained matrix. The deposits are also characteristically contained in large volumes of rock that have been shattered, sheared, faulted, or somehow broken up on a fine scale, and through which mineralizing fluids have found

FIGURE 4–6 *Aerial photograph of the Morenci Open Pit, from which the Phelps Dodge Corporation mines the porphyry copper deposit at Morenci, Arizona. The size of the pit can be gauged from the string of rail cars carrying ore, visible in the left center of the photo, and from buildings visible on the rim of the pit. (Courtesy Phelps Dodge Corporation.)*

easy passage. The shattering and mineralization may be in the intrusive, in the surrounding rocks, or both, and are apparently due to the violent and often volcanic forces associated with intrusion of the porphyritic rocks. Geologists are still unsure of all the events leading to formation of porphyry coppers, although they have occurred repeatedly in some areas, such as the copper belts of the Southwestern U.S. and of the western Andes in Peru and Chile (Fig. 4–7), both areas of widespread volcanism within the last 170 million years, during which all the known porphyry coppers formed.

The first porphyry copper to be recognized is still worked at Bingham Canyon, Utah. It is now the largest copper producer in the U.S.A., but D. C. Jackling and R. C. Gemmell were treated with skepticism when in 1899 they first proposed the idea—revolutionary at the time—of bulk mining low-grade ores; with financial backing, the idea proved successful. By 1907 a mill was built capable of handling the then huge tonnage of 6,000 tons per day. Capacity is now 125,000 tons! More than 25 porphyry coppers have now been discovered and mined; although principally clustered in the Americas, they have been discovered in the U.S.S.R. in the Kounradskiy area of Kazakhstan, in the Bor-Majdanpek area of Yugoslavia, and most recently in the Bougainville area of New Guinea.

A second type of large copper deposit, commonly called *stratiform* and believed by many to have formed as copper-rich sediments, accounts for approximately 25 per cent of the world's production of copper. Stratiform deposits—so called because they are confined to individual sedimentary horizons —are known from several geological ages and have a wide geographic distribution. Each stratiform deposit has features somewhat different from others in the class, but since copper-rich sediments are not known to be forming today, direct observation cannot provide answers to the many puzzles about their origins.

The longest worked and most famous stratiform deposit in the world, and the one least altered by metamorphism—hence, the most informative—was laid down in Permian times in the area now occupied by Central Europe. A sedimentary bed of organic-rich muds, rarely more than two feet thick, but now containing an usually high concentration of Cu, Pb, and Zn sulfides, formed over an area of 20,000 square miles in a shallow sea of the time called the Zechstein (Fig. 4–8). The mud, now consolidated to a shale and called the *Kupferschiefer*, was not uniformly mineralized, for the copper ores are strongly concentrated in small areas in the western part of the basin—in East Germany and Poland. It is not certain that the sulfides precipitated contemporaneously with the muds or whether they were introduced during *diagenesis*, the period immediately following burial of a sediment, when many chemical and physical changes occur in the enclosed organic and mineral matter. It has been observed, for example, that sulfide minerals replaced the

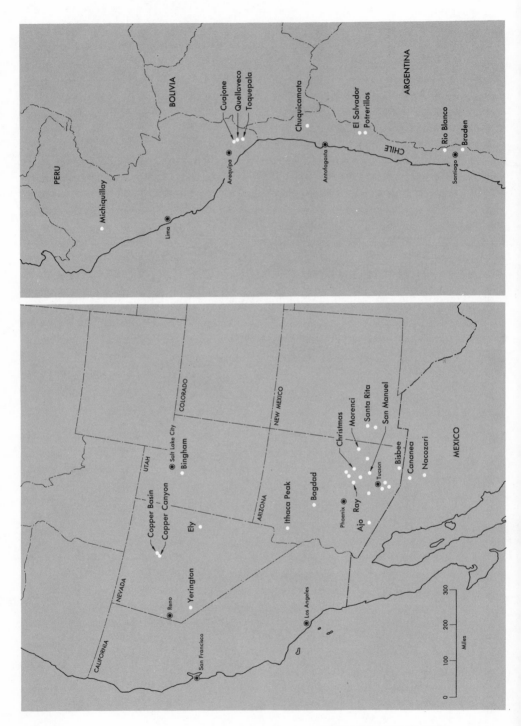

Metals: the scarce elements

FIGURE 4–7   (Opposite page) Porphyry copper deposits form two remarkable clusters, one in the southwest United States and the other in the western Andes of Peru and Chile.

cartilage of fossil fish, which indicates their post-sedimentary origin. However, delicate fossils like fish are often broken and distorted by the compaction that accompanies the earlier stages of diagenesis; in the Kupferschiefer, the breakage occurred after replacement of the cartilage, so introduction of the sulfides must have been very early.

The Kupferschiefer has been only slightly changed since deposition. There are also a number of large deposits in Precambrian sediments, and these have been more intensively metamorphosed, making evidence of their sedimentary origin ambiguous. In the United States, the only important deposits of the type —in the White Pine area of Michigan—

FIGURE 4–8   (Below) Extent of the shallow Zechstein Sea in which, during the Permian Period, the thin, metal-rich bed now called the Kupferschiefer was laid down. (After R. Brinckmann, 1960.)

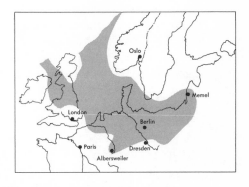

have many characteristics in common with the Kupferschiefer. Of considerable importance too are the deposits in the Dzhezhazgan-Karsakpay area of Kazakhstan, but the most remarkable of all occur in a belt of exceedingly rich deposits in Zambia and the Congo (Fig. 4–9), now known as the *Zambian-Katangan*

FIGURE 4–9   *The Zambian Copper Belt, a remarkable series of rich copper deposits in Zambia and the Congo, believed by many geologists to have been laid down as copper-rich sediments along a Precambrian shoreline. The deposits now produce about 20 per cent of the world's copper. (After F. Mendelsohn, 1961.)*

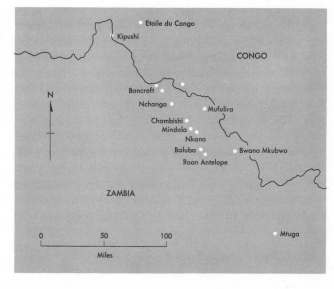

Copper Belt, but also commonly referred to as the Rhodesian Copper Belt.

We are faced with the puzzle of why one sediment is rich in sulfides whereas another one, seemingly identical, is not. The most direct answers to the problem came from two separate but quite important observations in 1962 and 1963. In the former year, a deep well drilled near the Salton Sea, in Southern California, tapped a hugh reservoir of hot (300°C–350°C) brine with remarkably high contents of scarce metals—concentrations almost 100 times higher than those of other known brines. During a three-month period, brine from one well deposited nearly ten tons of a sulfide-rich scale containing 20 per cent Cu and 8 per cent Ag. The first reaction was to suspect that the brines were hydrothermal solutions from a deep-seated magma, but studies of the isotopic composition of the water and its dissolved load have shown that this was not, after all, the case. The waters had their origin at the surface, were heated in depth, and then reacted with the minerals in the enclosing reservoir rocks. The reservoir rocks were recrystallized, and in the process the trace amounts of the scarce metals that they contained as substituting elements were released to the brine. The brines are therefore metamorphic hydrothermal solutions.

The second answer to the puzzle came from the other side of the world. Oceanographic studies in the Red Sea during 1963 disclosed three places where hot, metal-rich brines like those in Southern California currently debouch into sea-floor basins, with the metals being redeposited as sulfides in the sediments. Under certain conditions, therefore, hydrothermal solutions of either magmatic or metamorphic origin can apparently provide the metal source for localized sedimentary sulfide deposition, and this is believed by many to be the origin of the stratiform ores.

Approximately 50 per cent of the world's copper production comes from porphyry coppers, the largest share coming from the U.S.A., and an estimated 25 per cent from stratiform deposits, principally those in Zambia. The remainder comes from a great variety of small deposits that are widely distributed and that afford most countries at least a small production. Because of the overwhelming importance of the porphyry and stratiform deposits, however, seven countries account for fully 83 per cent of the world's production (Table 4–1).

Lead and Zinc

Lead and zinc are discussed together because their ore minerals occur together, and in each case a single mineral species, galena (PbS) and sphalerite (ZnS) respectively, accounts for most of the world's production. Lead is used principally for storage batteries, which in turn are used mainly in automobiles, and for lead tetraethyl, an antiknock additive to gasoline. Zinc, used principally as a component of brass until the eighteenth century, has many uses today, but more than 50 per cent of production is consumed in the preparation of alloys for die-cast products, and in anticorrosion treatment of iron and steel.

Table 4–1

Leading Copper-producing Countries, 1967, and Indicated Reserves at the End of 1965

Country	Production (Tons)	Reserves (Tons)
U.S.A.	954,000	32,500,000
U.S.S.R.	880,000	35,000,000
Chile	732,000	46,000,000
Zambia	730,000	25,000,000
Canada	603,000	8,400,000
Congo	353,000	20,000,000
Peru	200,000	12,500,000
Other countries	884,000	33,000,000
Total	5,436,000	212,400,000

After U.S. Bureau of Mines.

Like copper ores, lead and zinc deposits occur in two quite different ways: first, as hydrothermal or contact metamorphic deposits, and second, as sedimentary deposits. In both cases the ores tend to be rich but confined in size, so that costly underground mining procedures are necessary for recovery. Large, low-grade types of deposits analogous to the porphyry coppers have not been discovered so far.

Hydrothermal deposits are widespread and usually closely associated with igneous intrusions, but one particularly unusual variety is an exception to the rule; it is known as the *Mississippi Valley type*, after the remarkable metallogenic province stretching from Oklahoma and Missouri to southern Wisconsin (Fig. 4–10). Deposits of similar affinity have been identified in several parts of Europe, northern Africa, northern Australia, the U.S.S.R., and, most recently, at Pine Point in Canada's Northwest Territory. This last is quite possibly the largest deposit ever discovered, and may belong in the same metallogenic province as the deposits in the United States.

Mississippi Valley deposits occur principally as replacement bodies in limestones of many ages. Solutions carrying the scarce metals apparently dissolved the limestone and slowly deposited the galena and sphalerite that often

Metals: the scarce elements

FIGURE 4–10 *Lead- and zinc-rich deposits of the Mississippi Valley type occur in Paleozoic limestones, dolostones and shales within a remarkable metallogenic province crossing the central portion of the United States. The province is bounded on the east by the Appalachian Mountains and on the west by the Rocky Mountains. (After A. V. Heyl, 1968, Economic Geology, v. 63, p. 586, with permission of the publisher.)*

form large and beautiful crystals (Fig. 4–11). The deposits are usually far from any obvious igneous activity. A controversy of long standing surrounds the source of solutions depositing the ores. In recent years it has been established, from the analysis of tiny fluid samples trapped as inclusions in imperfections of the growing crystal (Fig. 4–12), that the depositing solutions were brines with close affinities to those from some oil fields as well as to the previously mentioned metamorphic hydrothermal brines found in the Salton Sea drilling. It has also been established that deposition occurred at 150°C or less, a relatively low temperature. From the evidence, many workers now feel that the slow escape of hydrothermal solutions of metamorphic rather than magmatic origin was responsible for deposition of the ores.

Metals: the scarce elements

FIGURE 4–11 *Large, well-formed crystals of galena (PbS) from the Mississippi Valley type deposits near Pitcher, Oklahoma suggest slow growth and absence of the frequent shattering and tectonic activity characteristically associated with many hydrothermal deposits. The cubes of galena in this photograph are 2 inches across.*

An increasingly large production of lead and zinc ores has come from deposits that many believe were originally sediments. The clearest example of a probable sedimentary deposit is the Kupferschiefer, where the deeper parts of the Zechstein basin apparently favored the accumulation of lead and zinc sulfides rather than of copper. Although the ores are low in grade, the Kupferschiefer has been worked for both lead and zinc in many places. Like the sedimentary copper deposits, most of the sedimentary lead-zinc ores are Precambrian and are now so changed by metamorphism that clear evidence of their origin has been obliterated. The two largest known deposits of the

FIGURE 4–12 *Inclusion of a saline ore fluid trapped in a crystal of quartz as it grew in a hydrothermal vein. The tiny inclusion, only two thousandths of an inch long, was filled with a homogeneous liquid at the moment of trapping, but under the lower temperatures at the surface of the Earth the liquid has cooled and contracted, forming the round vapor bubble on the right, and has deposited crystals of salts that were in solution at the higher temperatures. (Courtesy Edwin Roedder.)*

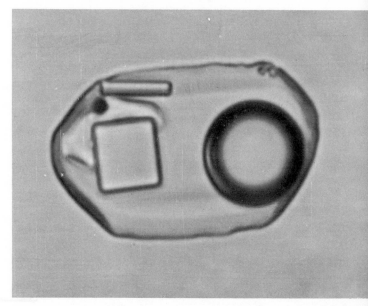

type are both in Australia, the first at Broken Hill, in the arid western part of New South Wales, and the second at Mt. Isa, several hundred miles to the north in Queensland.

The world's production of both lead and zinc is strongly dominated by four countries (Table 4–2).

Table 4–2

Leading Lead- and Zinc-producing Countries*

Country	Lead (Tons)	Percentage of World Production	Zinc (Tons)	Percentage of World Production
U.S.S.R.	440,000	14.0	590,000	11.4
Australia	417,000	13.3	446,000	8.6
Canada	340,000	10.9	1,249,000	24.1
U.S.A.	317,000	10.1	549,000	10.6
World Total	3,133,000		5,175,000	
Indicated World Reserve	50,000,000		85,000,000	

After U.S. Bureau of Mines.
* Using mine production figures for 1967.

Nickel

Nickel, used almost entirely as an alloying metal in the production of special products such as stainless steel and high-temperature and electrical alloys, is a product of twentieth-century technology. Smelting and working the metal is so difficult that old German miners, who confused the similar-looking copper and nickel sulfide ores, called it kupfernickel after the infernal "Old Nick," who supposedly bewitched the copper (kupfer) ore and thus made it impossible to handle. The frustrations of these miners live on in our use of the word nickel. Owing to modern technology, the world production of nickel rose in 1966 to 481,000 tons (Table 4–3), of which the U.S., though a minor producer, consumed 174,000.

Table 4–3

Leading
Nickel-producing Countries*

Country	Nickel (Tons)	Percentage of World Production
Canada	247,000	51
U.S.S.R.	105,000	22
New Caledonia	68,000	14
Cuba	26,000	5
World total	481,000	

After U. S. Bureau of Mines.
* Using 1966 production figures.

Metals: the scarce elements

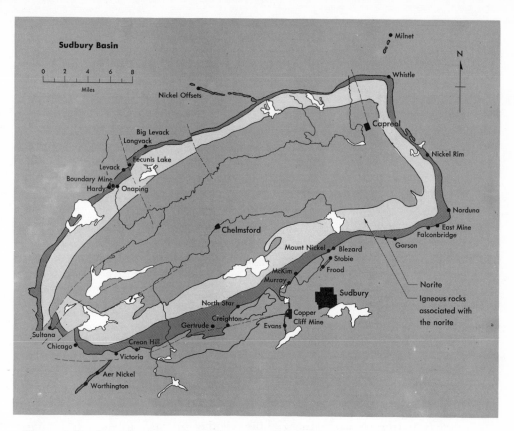

FIGURE 4–13 *Location of the rich nickel and copper ore bodies near the base of the basin-shaped intrusion of norite, a distinctive mafic rock, near Sudbury in Central Ontario. It is believed that molten nickel and copper sulfide liquids separated from the cooling mafic magma. The sulfide liquid, being dense, sank to the bottom of the magma chamber and accumulated. As the complex cooled, the sulfide ores that we see today formed either from crystallization of the sulfide-rich puddles or from cooling of sulfide-rich liquids that were squeezed into fractures in the rocks underlying the norite intrusive by tectonic movements and other geologic forces. (After J. E. Hawley, 1962, Can. Min., v. 7)*

There are only two important ways in which nickel is recovered. The first, accounting for the large Canadian production (Table 4–3), is from the mineral pentlandite, which commonly occurs in deposits formed by magmatic segregation. Just as some cooling magmas become saturated in molten iron oxide and form a new liquid that is immiscible with the host, so others become saturated in molten sulfides. The circumstances that bring this about are rare and are known to occur only in mafic magmas. The dense, immiscible sulfide liquid sinks to the bottom of the magma chamber, accumulating as sulfide-rich puddles (Fig. 4–13) which later crystallize to the complex mixtures of Ni, Cu, and Fe sulfides that are mined at the great Sudbury, Ontario, and Thompson Lake, Manitoba, deposits—the two largest known in the world. Though

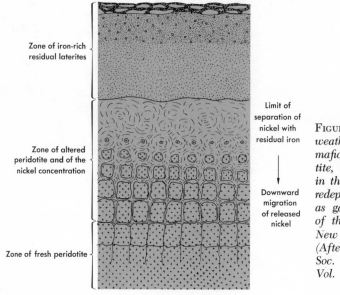

Surface

Zone of iron-rich
residual laterites

Zone of altered
peridotite and of the
nickel concentration

Zone of fresh peridotite

Limit of
separation of
nickel with
residual iron

Downward
migration
of released
nickel

FIGURE 4–14   *Chemical
weathering of nickeliferous
mafic rocks, such as peridotite, releases nickel trapped
in the olivine which is then
redeposited as minerals such
as garnierite. Residual ores
of this kind are worked in
New Caledonia and Cuba.
(After E. de Chétalat, Bull.
Soc. Geol. France, Sér. 6e,
Vol. XVII, 1967, p. 129,
Fig. 4.)*

much controversy has surrounded the origin of the Sudbury deposits, leading
to suggestions of a hydrothermal origin for the ores, recent work has rein-
forced the conclusion that the ores are indeed magmatic segregates.

The second locale of nickel is in the residual weathering zones formed over
many mafic rocks in tropical regions. We previously used the example of
nickel substitution in an olivine. Weathering releases the trapped nickel to
surface waters, and under some conditions it is reprecipitated as relatively
insoluble nickel silicates (Fig. 4–14), such as garnierite, named for the French-
man Garnier who first discovered the rich nickel ores of this type in New
Caledonia. Beneficiation, which is so costly to man, has thus been effected by
nature in a slow, long-continued weathering process.

Although quite pure deposits of garnierite may be formed in this fashion,
it is more common for tropical weathering of mafic rocks to produce an iron-
rich laterite in which the concentration only reaches a level of about 1 per cent
Ni. Low-grade deposits of this kind are widespread in the tropics, and nickeliferous laterites measured in the hundreds of millions of tons have been
discovered in Cuba, the Philippines, Greece, Borneo, and other parts of the
world. Though only worked in Cuba at the present time, the laterites constitute man's largest known potential resource of nickel, and together with
the large Canadian deposits they appear to be adequate to meet our demand
for nickel far into the future.

Molybdenum

Molybdenum, like nickel, is a metal of twentieth-century technology. Used mainly as an alloying element in steels, to which it imparts toughness and resilience, molybdenum first gained widespread importance when used in steels for armor-plate and armor-piercing shells during World War I. It now has a wide variety of uses, particularly in alloys where resistance to wear and retention of strength at high temperatures are required.

FIGURE 4–15 *Distribution of disseminated molybdenum deposits in a metallogenic province that falls near the eastern margin of the Cordillera of North America. (After K. F. Clark, 1968, Economic Geology, v. 63, p. 560.)*

One mineral, molybdenite (MoS_2), accounts for most of the world's production of molybdenum, but because it has a highly erratic distribution it is difficult to estimate accurately the world's resources. A significant fraction —approximately 25 per cent of the world's current molybdenum production—is derived as a by-product of porphyry copper mining. The molybdenum content of porphyry copper cores is low—in the range 0.01 to 0.04 per cent Mo—but the tonnages of ore processed are so large, and the costs of concentrating molybdenite by flotation so low, that a profitable recovery is effected at many of the larger porphyry copper mining operations.

The site of the major production and future resources of molybdenum is a series of deposits that have many geological similarities to porphyry coppers. Sometimes called the porphyry molybdenums, but more commonly called *disseminated molybdenums* (because the ore minerals are finely disseminated through a large volume of rock), the deposits are known to lie in a metallogenic province stretching from the Mexican border to northern British Columbia (Fig. 4–15). Analogous to the porphyry copper deposits, the disseminated molybdenums are associated

with shallow, silicic igneous intrusives; are deposited in intensely broken ground; and, because of their large volumes, are amenable to low-cost, bulk mining practices. One of the disseminated molybdenums, at Climax, Colorado, has supplied at least half of the world's production for the last 50 years. Such heavy dependence on a single ore deposit is unwise, of course, and the situation is now being eased by the commencement of large-scale mining operations at other deposits in Canada and the United States.

Although accurate molybdenum production figures are not available for communist-bloc countries, it is estimated that current world production rates are approximately 60,000 tons per annum. With indicated reserves of 900,000 tons of moybdenum from Climax alone, and with even larger reserves reported for the Henderson deposit, we are unlikely to face a resource problem with molybdenum supplies for many years to come.

Silver

The widespread attention paid to silver shortages by newspapers and other news media is derived from its use in currency and common household silver plate. The shortage arises from industrial consumption, however, with the photographic and electrical industries consuming an ever larger share of a supply that cannot be easily expanded to satisfy demands.

Shortly after Columbus blazed a pathway to the Americas, the great silver bonanzas of Central and South America were discovered. To the present day, deposits in the Cordilleran chain, stretching from Alaska to Tierra del Fuego, have remained the major source of the world's silver supply. Approximately 55 per cent of the silver produced in 1967, for example, came from countries in the Americas (Table 4–4), with most of the remainder coming from Australia and the Soviet Union, and only marginal production from Asia and Africa.

Silver minerals commonly occur in hydrothermal vein deposits and, characteristically, either they are associated with lead, zinc, and copper minerals, or else the silver itself is carried in the lead and copper minerals by atomic substitution. Only a minor percentage of silver-producing deposits are rich enough to be worked for silver alone, however, and therein lies the silver

Table 4–4

Leading
Silver-producing Countries*

Country	Silver (Tons)	Percentage of World Production
Mexico	1310	14.6
Canada	1250	13.9
Peru	1230	13.7
U.S.S.R.	1200	13.4
U.S.A.	1100	12.3
Australia	680	7.6
World Total	8970	

After U. S. Bureau of Mines.
* Using 1967 Production Figures.

Metals: the scarce elements

problem. Most silver is produced as a valuable by-product from copper-, lead-, and zinc-mining, and its production rate is controlled strictly by the production rates of the associated metals. Though a few new silver deposits have been found in recent years, so much silver is produced as a by-product that silver production has not been able to expand to meet the growing need. With the 1965 decision by the United States Government to withdraw silver from currency, the price of silver has climbed steadily, and it may be anticipated that further increases are ahead unless large new supplies are found. Such discoveries do not seem likely, however. Though old mines may be reactivated and formerly uneconomic ores become profitable in the face of rising silver prices, there seems little hope that silver production will grow to meet all demands; rather, it is likely that future uses will have to be curtailed to meet the limited supply.

Other Sulfide-forming Scarce Metals

The remaining scarce metals that occur principally as sulfides do not warrant separate discussion. Almost without exception, small deposits are widespread and resources are more than adequate for foreseeable future needs. The one apparent exception is mercury.

Most of the world's mercury production comes from one mineral, the vermillion-colored cinnabar (HgS), which is found erratically distributed in narrow hydrothermal veins in a number of volcanic areas. The known deposits are all shallow, and most are so small that they have been exhausted soon after discovery. The world's current production comes largely from Spain—where the Almadén mine has been producing for more than 2,000 years—and from Italy, both regions of extensive Cenozoic volcanism. The once great deposits of California and Nevada are apparently almost exhausted. Known reserves of mercury are small and its by-production from other mining activities limited. Many experts now suspect that it will be the first scarce metal to exhaust available supplies, and some believe this day will be reached in the present century.

Scarce Metals
Commonly Occurring in the Native State

The scarce metals occurring in the native state (Table 4–5) are a less diverse group than the sulfides. Platinum, palladium, rhodium, iridium, ruthenium, and osmium, collectively called the platinoid elements, always occur together and are concentrated in mafic igneous rocks. The abundance levels of the platinoids are all low—platinum and palladium, the most abundant of the group, are only present in the crust to an extent of 0.0000005 per cent; the others are even less abundant. However, the platinoid abundance in mafic rocks derived from the mantle is noticeably higher, though not sufficiently so for the mafic rocks to be considered potential resources of platinoids without additional concentrating factors coming into play.

Table 4-5

Leading Producers of the Scarce Metals
Commonly Occurring in the Native State*

Country	Platinoids (Tons)	Percentage of World Total	Gold (Tons)	Percentage of World Total
U.S.S.R.	65.4	60.3	196	12.5
Republic of S. Africa	28.5	26.3	1050	66.9
Canada	13.9	12.8	102	6.5
U.S.A.	0.6	0.5	54	3.4
Australia	—	—	22	1.4
World total	108.5		1569	

After U. S. Bureau of Mines, 1967.
* The platinoid elements Pt, Pd, Rh, Ir, Ru, and Os have been grouped together because of their similar occurrences and uses. Within the group Pt and Pd each account for about 40 per cent of the production, Rh 9 per cent, Ir 6 per cent, Ru 4 per cent, and Os 1 per cent.

Immiscible sulfide liquids preferentially concentrate the platinoid elements with respect to the parent magma. In the mining of these sulfide segregation deposits for nickel and copper, the platinoids are recovered as valuable by-products. The largest mafic intrusive so far discovered in the world, the Bushveldt Complex of South Africa, contains a zone called the Merensky Reef (Fig. 4–16) in which both sulfides and platinoids are greatly enriched; this single indicated reserve is enough to assure the world's supply of platinoids for centuries. There is another important occurrence of platinoids, however. The metals are essentially unaffected by chemical weathering and are malleable, so grains do not disintegrate as they are transported and deposited in sediments; being very dense, they concentrate in placers (Fig. 4–17). Where the sediment source area contains many basic igneous rocks, as in the Ural Mountains in the Soviet Union, important placer concentrations of platinoids develop.

Gold, unlike the platinoids, is not commonly associated with mafic rocks, but is characteristically associated with silicic igneous rocks. Like the sulfide minerals, gold is apparently transported by magmatic hydrothermal solutions,

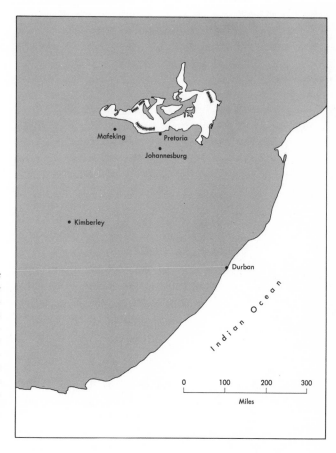

FIGURE 4–16 *One of the largest igneous complexes ever discovered, the Bushveldt Complex of South Africa, contains a platinoid-rich zone called the Merensky Reef in the lower, mafic portion of the complex. Here are found the world's largest indicated reserves of platinoids. (After R. A. Pelletier, 1964, Oxford University Press, Cape Town.)*

though how this is done remains a point of conjecture, for native gold is one of the least soluble substances known; it is commonly found in hydrothermal vein deposits, either associated with sulfide minerals or alone. Very rarely does an ore contain much gold; a rich ore, for example, contains only 0.007 per cent Au, and yields about 2 ounces of gold for every ton of rock mined. Much leaner ores than this can now be profitably worked.

Gold, like the platinoids, is resistant to most forms of corrosion and is almost indestructible. Most of the gold ever mined is still available, having been used and reused many times in its passage through history. Some of the gold from Cleopatra's bracelets, for example, may reside in the tooth fillings or wedding ring of a present-day housewife. Gold is also an unusual scarce metal in that it forms a separate mineral even at very low concentration levels and is apparently not concealed by atomic substitution to the same extent

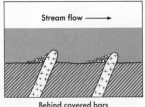

Behind covered bars

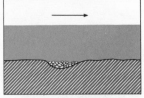

In covered rock holes

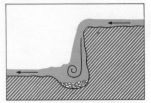

In potholes below waterfalls

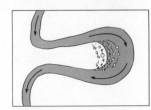

On the inside of meander loops

FIGURE 4–17 *Typical sites for placer accumulations which occur where obstructing or deflecting barriers allow faster-moving waters to carry away the suspended load of light and fine-grained material while trapping the more dense and coarse particles which are moving along the bottom by rolling or only partial suspension. Placers may form wherever moving water occurs, though they are most commonly associated with streams.*

Downstream from the
mouth of a tributary

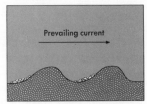

In the ocean behind bars against
the prevailing current

that many other, even more abundant, elements are. Because gold is widespread in tiny amounts and is very dense and indestructable, it is ideally suited for concentration in placer deposits, from which has come much of the world's production (Fig. 4–18).

Among the most remarkable of all the mineral deposits in the world are the gold deposits of the Witwatersrand district of South Africa. Formed in Precambrian times, the Witwatersrand ores are contained in a series of conglomerates—sedimentary rocks consisting of rounded pebbles cemented by a finer-grained matrix. This is exactly the rock type that commonly carries placer concentrations in most parts of the world and in all geologic ages; there the story would end but for the astonishing extent of the Rand ores. Most placer deposits are small, perhaps a few hundred or thousands of yards in extent, and are clearly confined to present or former narrow stream channels. The Rand ores too are apparently concentrated in what were old channels, but the deposits have been mined along an outcrop length of conglomerates in excess of 250 miles (Fig. 4–19), and they have been followed to the limits of practical mining—more than two miles deep. Although most who have studied these are hesitant to propose alternatives to placer deposition for the Rand ores, no one has successfully solved the problem of their extent, nor of their apparent uniqueness.

FIGURE 4–18 *A large floating dredge recovering gold from stream placers near Fairbanks, Alaska. The front of the dredge is on the right. A line of dredge buckets scoops the gravels up from the bottom of the pond. The gold is separated in the dredge and the gravels are deposited behind the dredge by the long boom on the left. The distinctive waste piles are formed by the swinging of the boom. (Photograph by Bradford Washburn.)*

First discovered in 1885, the Rand district soon became, and has remained, the world's leading gold-producer. In 1968 it accounted for more than 60 per cent of the world's total production. Although gold production is widespread in small amounts around the world—71 countries have recorded some production—large deposits are rare, and only five countries account for 90 per cent of production.

Scarce Metals
Forming Oxygen-containing Compounds

With the exception of tin, the scarce metals with affinities for oxygen are all newcomers to man's technological array, and the diverse uses to which they are put reflect the amazingly complex technology we now support. Tungsten, tantalum, vanadium, and niobium (formerly known as columbium) are principally used as alloying agents in special steels but also have other highly specialized uses. For example, 33 per cent of the tungsten produced is now used

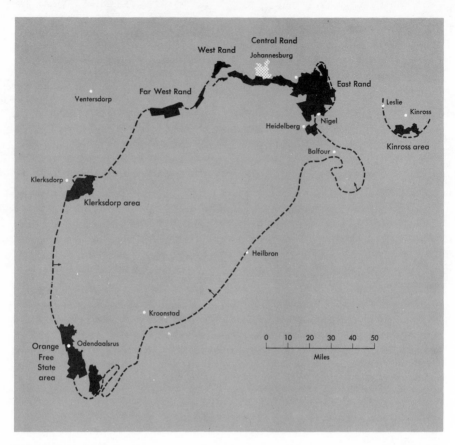

FIGURE 4-19 *Gold has been mined from the remarkably continuous gold-bearing con-glomerates of the Witwatersrand System along a strip more than 250 miles in length. Most of the gold is situated in beds of the Upper Division that dip inwards (direction of arrows) to form a natural basin-shaped structure. Most of the basin is too deep for mining, and much of it is now covered by later rocks. The areas that have actually been mined are shown in black. (After R. A. Pelletier, 1964, Oxford University Press, Cape Town.)*

in the manufacture of tungsten carbide (WC), an extremely hard substance used for cutting-edges in metal work, in rock drills, and in armor-piercing shells, while tantalum has been proved to have the special electronic properties needed for low-power valves, and offers revolutionary possibilities for computers of the future. Uranium, of course, has become the wonder metal of our age now that atomic energy can be released by controlled fission. Because its uses are those of a fuel rather than a metal, uranium will be separately discussed in Chapter 7.

Tin, the only scarce metal of the group with a large production tonnage, has been used as an alloying agent in bronze for thousands of years, but the largest use is now in the tinplate industry, where it serves as an anticorrosion coating on iron and steel, and in the production of soft solders.

Tin and tungsten minerals form in the same geologic environment and often occur together. They are found concentrated adjacent to siliceous igneous rocks either as contact metamorphic deposits or as the first minerals deposited in magmatic hydrothermal veins. Their most common minerals—cassiterite (SnO_2), wolframite ($FeWO_4$), and scheelite ($CaWO_4$)—are dense, erosion-resistant minerals that are readily concentrated in placer deposits. Much of the world's supply has come from such secondary concentrations.

Tin and tungsten are both elements for which the problems of future resources loom large; their abundances are low and their deposits scarce. North America, for example, is almost devoid of known tin deposits, though it is more fortunate in its tungsten resources. Most of the world's tin production and known reserves are concentrated in two narrow metallogenic belts, one that runs first along the Malayan Peninsula, then southeast to Java and to Indonesia, and another that runs along the eastern side of the high Andes, in Bolivia and Peru. The major tungsten production and reserves are also restricted geographically, with approximately half of the world's production coming from an east-Asian belt running from Korea, through southern China, to Malaya.

Niobium, tantalum, and beryllium are all found concentrated in a special, rare type of igneous rock called a *pegmatite*. Pegmatites are small—rarely more than a few dozen feet in their longest dimension—and are associated with siliceous igneous rocks crystallized at great depth. It is believed that they are formed by the water-saturated residue left after crystallization of the bulk of the magma, and that they only form under deep-seated, high-pressure conditions where water and other volatiles cannot readily escape to form hydrothermal deposits. Elements such as niobium and beryllium, which do not readily substitute in the earlier-formed silicate minerals, become concentrated in pegmatites. Because the pegmatites form at considerable depth in the crust, they are only exposed in deeply eroded terrains.

In recent years large deposits of niobium and tantalum minerals have been found in a unusual igneous rock called *carbonatite,* the origin of which also remains somewhat of a mystery. Carbonatites are intrusive rocks composed largely of calcium carbonate. Most rocks rich in calcium carbonate are sediments formed at the Earth's surface that contain a distinctive array of minor elements. Carbonatites, however, contain an unusual and quite different array of minor elements, including niobium and tantalum; this strongly suggests that they are not generated by any melting process involving deeply buried carbonate sediments. Rather, they seem to be derived by some presently unknown means of differentiation from other igneous rocks.

5

Industrial minerals:
chemical and fertilizer supplies

. . . the use of chemical fertilizer in 1960 was about six times the level of the 1930's. The additional production flowing from the recent rate of consumption of some 6 to 7 million tons of plant nutrients per year . . . is roughly equivalent to the yield from some 75 million acres of un-fertilized farmland. (H. H. Landsberg, Natural Resources for U.S. Growth, 1964.)

The nearly equivalent terms *industrial minerals* and *non-metallic minerals* are somewhat ambiguous, for they are neither strict scientific nor exact economic terms. They are widely employed, however, and embrace a group of minerals used for purposes other than those served by the metallic properties of the elements they may contain. Unlike the metals, the nonmetallic minerals cannot be readily classified in terms of crustal abundances; they can, however, be simply classified on the basis of use. First, there are the minerals of primary use for fertilizers and for raw chemicals, accounting for 33 per cent of the value of the nonmetallic production. Second, there are the materials for the building and construction industries discussed in Chapter 6.

Minerals for fertilizers

The fertilizer minerals are one of our most vital resource needs, essential to increased food production for the expanding human population. Fortunately, most of the necessary compounds are abundant and widely distributed, and we seem to be in no danger of exhausting the supplies.

Plant growth requires many elements. Oxygen and hydrogen (both derived from water) together with carbon (derived from atmospheric carbon dioxide) make up 98 per cent of the bulk of the living plant. But nitrogen, phosphorus, potassium, calcium, and sulfur are also essential, and for the land plants—the source of our food supplies—they are provided by the soil. The rate of supply in part determines the rate of plant growth, and, to have any effect at all, the elements must be supplied in a form that the plant can assimilate. For example, most soils contain 1 per cent or more of potassium, but to the plant, which needs soluble material, it is unavailable because it is locked in insoluble alumino-silicate minerals such as feldspars. Therefore, to enhance growth rates by addition of a potassum fertilizer, we usually add the potassium as the chloride (KCl), the most abundant soluble potassium mineral. The efficacy of the nitrogenous, phosphatic, and other fertilizers similarly depends on their solubility, and it is in the form of soluble compounds, or compounds that can be rendered soluble by minimal treatment, that fertilizer resources are sought.

Potassium Fertilizers

Potassium is one of the triad of *essential fertilizers,* the others being phosphorus and nitrogen. With increasing food demands, it is not surprising that the world's consumption of potassium salts has been increasing almost 10 per cent a year.

Potassium itself is an abundant element, widely distributed in silicate minerals. It is the rarely formed soluble minerals that we seek as resources, however, and these are almost solely confined to a special class of mineral deposits, known as *marine evaporites,* that result from the accumulation of salts by evaporation of sea water. Because marine evaporites are also of vital importance in the production of other minerals, we will briefly discuss their origin.

The major elements dissolved in sea water were displayed in Fig. 2–4. They can be recast into the constituents that precipitate from sea water by balancing the positively charged cations, such as sodium (Na^{+1}), against the negatively charged anions, such as chlorine (Cl^{-1}), to preserve electrical neutrality (Table 5–1). Sodium chloride is the most abundant constituent; next follow the magnesium salts, then calcium sulfate and potassium chloride. When water is removed by evaporation, the remaining brine becomes increasingly concentrated until it finally reaches saturation, first in one salt, then in others. It is not necessarily the most abundant compound that precipitates first; saturation of the relatively insoluble—and therefore sparse—compounds is usually reached long before saturation of those that are highly soluble.

Table 5-1

Major Constituents
of Sea Water*

Constituent	Percentage of Total Dissolved Solids
NaCl	78.04
$MgCl_2$	9.21
$MgSO_4$	6.53
$CaSO_4$	3.48
KCl	2.11
$CaCO_3$	0.33
$MgBr_2$	0.25
$SrSO_4$	0.05

* Obtained by recasting the data in Fig. 2–4 into molecular compositions.

The first compound to precipitate from evaporating sea water is $CaCO_3$, for which the solubility is extremely low and the amount in solution small relative to NaCl. The next phase, $CaSO_4$, does not begin to separate until the solution has been reduced to 19 per cent of the original volume, and NaCl, the third phase to separate, only does so when the residual solution is reduced to 9.5 per cent of the initial volume. Precipitation of NaCl, plus a small amount of $CaSO_4$, continues until the brine is reduced to about 4 per cent of its original volume; then the first phase to contain either magnesium or potassium, a complex salt called polyhalite ($K_2SO_4 \cdot MgSO_4 \cdot 2CaSO_4 \cdot 2H_2O$), begins to precipitate. The amount of NaCl in solution is large to begin with, and considerably more than half of it will be precipitated during the reduction in solution volume from 9.5 per cent to 4 per cent, so the thickest layer formed during a single evaporation cycle will be the NaCl layer. The sequence of minerals separating from the final 4 per cent of the brine (called the bitterns) is complex and variable, depending on such factors as the temperature and whether the final liquid remains in contact, and hence can react, with the earlier-formed crystals. Two of the precipitates found in most sequences are sylvite (KCl) and carnallite ($KCl \cdot MgCl_2 \cdot 6H_2O$), and it is in these late-stage evaporate minerals that most of the world's useful resources of potassium minerals are to be found.

The evaporation of a completely isolated body of sea water would produce the sequence and volume of salts shown in Fig. 5–1. When we examine actual evaporite deposits, these volume relationships are rarely found—the early-formed precipitates, $CaCO_3$ and $CaSO_4$, are much more abundant, and the late-formed precipitates, the K and Mg salts, are rare. Furthermore, complete evaporation of a body of sea water even as deep as the Mediterranean, which averages about 4,500 feet, would produce only 78 feet of halite and

Industrial minerals: chemical and fertilizer supplies

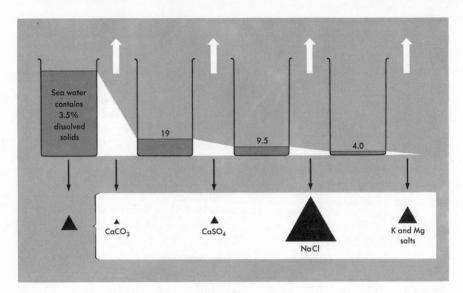

FIGURE 5-1 *Succession of compounds precipitating from sea water. When evaporation reduces the volume to 19 per cent, $CaSO_4$ begins to precipitate; at 9.5 per cent, NaCl; and so on (expressed in percentage.)*

layers of gypsum only 4.5 feet thick. However, beds of $CaSO_4$ and NaCl many hundreds of feet thick are known geologically. Clearly, some mechanism other than evaporation of a totally isolated basin must occur. The common circumstance is to find precipitation in a partially isolated basin from which water is removed by evaporation but into which fresh sea water is continually flowing. There are several geologic circumstances where such an arrangement occurs, and they are collectively called barred basins. Water flows into the basin over a submerged bar; evaporation of the surface waters continually enriches the basin in dissolved salts because the partially enriched but heavier brine sinks to the bottom and is prevented from recirculating by the restricting bar. In this fashion the salinity of the basin increases more slowly than it would by direct evaporation of an isolated body, and the brine remains for a long time in the salinity range where $CaCO_3$ precipitates. It is even possible for the precipitated $CaCO_3$ completely to fill the basin to the level of the bar before a sufficiently high salinity for the precipitation of $CaSO_4$ is reached. Similarly, thick beds of $CaSO_4$ may form and the brines never reach the NaCl stage. The frequency of occurrence *and* the total thickness of evaporite salts in sedimentary basins around the world *decrease* in the order $CaCO_3 > CaSO_4 > NaCl > $ K and Mg salts (Fig. 5-2).

At the present time evaporite deposits are accumulating on every continent. Although high temperatures are not essential for concentration—drying winds can accomplish the same thing in cold climates—the evaporites are formed mostly within a belt 35° of latitude north and south of the equator. The modern deposits are all small, and there is not at present any large marginal sea with a restricting flow—such as the Baltic, the Mediterranean, or the Black Sea—that fulfills both the morphological and the necessary climatic conditions of a large evaporite deposit. As we look at the geological record, however, we find that this is a temporary situation, for marine evaporites are widely spread both in time and in space, and there have been several times in the past when world-wide conditions, such as prolonged periods of higher temperatures, were much more conducive to the formation of marine evaporites than they are at present. Thus, during the Permian Period, exceptionally thick evaporite sequences were formed in North America and Europe.

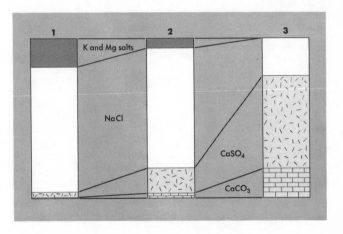

FIGURE 5–2 *Comparison of precipitation profiles from (1) the single-stage evaporation of sea water, (2) the rich Zechstein evaporite section in Germany, and (3) the most commonly observed evaporite section, where the potassium salt stage is not reached. (From* Salt Deposits, *by H. Borchert & R. O. Muir, Copyright © 1964, by Litton Educational Publishing, Inc., by permission of Van Nostrand Reinhold Company.)*

It is somewhat ironic that the circumstances of World War I, which closed the Permian potassium deposits of Germany to the United States, should have been responsible for providing the impetus to geologists to find, and later develop, even richer deposits of the same type and age in New Mexico. A large, shallow Permian sea deposited a thick section of evaporite salts over an area of at least 100,000 square miles in what is now New Mexico, Texas, Oklahoma, and Kansas. In a portion of the basin, near Carlsbad, New Mexico, an estimated 3,000 square miles of the sequence contain potassium salts in beds that reach 12 feet in thickness. These deposits, which are among the richest in the world but small by comparison with several others, have indicated reserves of potassium salts in excess of 100 million tons.

North America has other large potassium reserves. A basin of Pennsylvanian age in southeastern Utah and southwestern Colorado, called the Paradox Basin, contains an estimated 7,800 square miles of potassium-rich salines, though much of it is too deep to warrant present recovery. In Saskatchewan, Canada, a huge and as yet incompletely explored resource of potassium salts has been

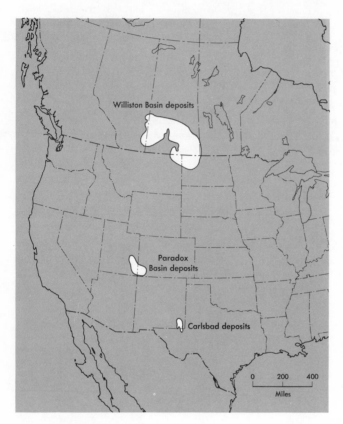

FIGURE 5–3 *Marine evaporite beds containing rich potassium resources are known to underlie large areas in North America. The reserves in these beds alone amount to more than 25 per cent of the world's reserves and are sufficient to supply world needs for many centuries. (After P. B. King, 1942, R. J. Hite, 1961, and S. R. L. Harding and H. A. Gorrell, 1967.)*

found in the Williston Basin (Devonian Period). Estimates of as much as 17,500 million tons of ore that are accessible by today's mining standards have been published, and they include only a small portion of the known basin outline (Fig. 5–3).

The Permian Zechstein Basin in Europe, previously discussed in connection with copper deposits (Fig. 4–8), also contains one of the world's greatest saline sequences. Where shallow enough to work, as in Germany and England, the basin has been a prolific producer of potassium salts, and much remains to be won from it. In 1960, Ruhlman reported indicated reserves in the East German portion of the basin to be 14 billion tons of potassium salts, and, in the deeper West German section, 20 billion tons of potential resources.

The world's resources of potassium salts are thus enormous, with large reserves in Italy, Spain, France, Israel, and the U.S.S.R., in addition to those discussed. Therefore, great expansions beyond even today's production rates can be supported for many centuries to come.

Table 5–2

Leading Producers of Potassium, 1967*

Country	Production, K_2O (Tons)	Percentage
United States	3,299,000	19.6
U.S.S.R.	3,040,000	18.0
West Germany	2,500,000	14.8
Canada	2,432,000	14.4
East Germany	2,400,000	14.2
France	1,962,000	11.6
World total	16,861,000	

* Because several different potassium salts are mined, it is convenient to express production in terms of the equivalent amount of oxide, K_2O.

Phosphorus Fertilizers

The phosphorous cycle appears to be the one most disrupted by intensive land cultivation. Phosphorus is a vital element for many cellular processes in the body, but man's major store of it is in his skeleton, composed of the mineral *apatite* ($Ca_5(PO_4)_3OH$), and this supply is not returned to the soils of our crop lands after death; it is therefore necessary to return it as a fertilizer. The only important source of phosphorus, and the only common phosphate mineral, is apatite; however, it is relatively insoluble. It is common practice, therefore, to treat apatite (or *phosphate rock*, as most of the apatite-rich raw materials are called) with dilute acids, usually sulfuric (H_2SO_4), to produce a more soluble material. The chemical reactions that ensue are complex, but the result, known as *superphosphate*, contains a high percentage of water-soluble compounds such as $Ca(H_2PO_4)_2$.

Apatite is widespread in trace amounts in most rocks, regardless of whether they are igneous, metamorphic, or sedimentary. Major concentrations of apatite in igneous rocks are known, but are comparatively rare; one such concentration, on the Kola Peninsula in the northern Soviet Union, is mined for its apatite content and is a major supplier of Soviet phosphatic fertilizers. The major production of apatite, however, accounting for more than 90 per cent of the present world production and for almost all of our known reserves, is from marine sedimentary deposits.

The origin of the marine deposits is still in doubt; as with other geologic problems, we cannot solve it by observing similar beds forming today, since none is known. In general outline, the process is believed to arise in the following manner. The waters of the ocean are near saturation with respect to apatite, and several effects can apparently cause saturation to be exceeded and precipitation to ensue. For example, nearly oxygen-deficient, or anaerobic, waters provide environments in which the water composition is considerably less alkaline than normal sea water and in which the solubility of apatite is exceeded. These environments are not common in the sea, especially on a large scale, and when they do form, the basin must be so situated that very little detritus is introduced to dilute the slowly precipitating apatitic sediments.

A particularly large and apparently unique deposit of this type was formed during Permian times in a shallow marine basin covering what are now portions of the states of Idaho, Nevada, Utah, Colorado, Wyoming, and Montana in the United States (Fig. 5–4). The phosphate-rich sediments, called the Phosphoria Formation, cover more than 100,000 square miles and reach thicknesses of 450 feet. Over most of the area the thickness of the phosphatic bed is only 3 feet or less, and at best it can be considered only a potential reserve. The tonnage, however, is enormous.

FIGURE 5–4 *Extensive phosphate deposits are worked in three important areas of the United States. The largest reserves are in the Phosphoria Formation and appear to be adequate for many centuries, provided that mining and transportation problems can be solved. (After V. E. McKelvey et al., 1953; Tenn. Div. of Mines, 1938; U.S. Geol. Surv. Bull., 1942.)*

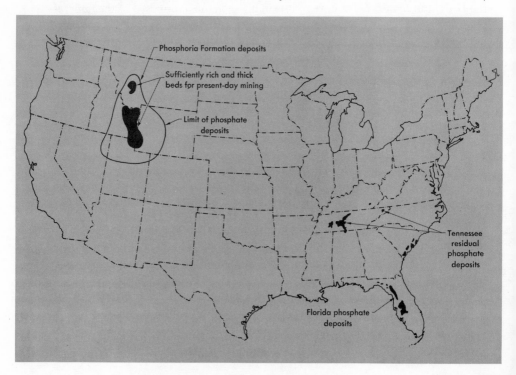

Small nodular bodies of precipitated apatite are common in some limestones formed in shallow marine water. Large concentrations of nodules, whether formed directly or concentrated secondarily into gravel beds, form valuable phosphate deposits. The huge "land pebble" phosphate deposits of Florida, source of most of the present phosphate production in the United States, are secondary concentrations of phosphatic nodules. Another large producing area, in Tennessee, contains residual deposits of phosphates formed by the weathering of nodule-containing limestones.

The principal phosphate-producing countries of the world are also the principal reserve-holders, and the indicated reserves are indeed huge (Table 5–3), assuring adequate supplies far into the future, even with large increases in consumption rates. The spotty geographic location of known resources is not encouraging, however, for transportation costs add heavily to the expense of the fertilizer. Countries such as Australia, with large land areas needing phosphatic fertilizers, must therefore actively pursue local exploration programs in the hope of finding accessible supplies.

Table 5–3

Major Phosphate Rock Producers, 1966*

Country	Production (Tons)	Indicated Reserves (Tons)
United States	39,050,000	14,500,000,000
U.S.S.R.	32,190,000	7,600,000,000
Morocco	10,405,000	21,000,000,000
Tunisia	3,527,000	2,000,000,000

U.S. Bureau of Mines
* Also owners of the largest indicated reserves.

Nitrogen Fertilizers

As mentioned in Chapter 2, nitrogen is the principal resource won from the atmosphere. The form in which it is added to the soil is either as a soluble nitrate, such as KNO_3 or $NaNO_3$, or as an ammonia compound, such as $(NH_4)_2SO_4$ or NH_4NO_3.

Most synthetic nitrogen compounds are produced either by one of the variants of the Haber-Bosch process—in which nitrogen from the atmosphere and hydrogen are combined under high temperatures and pressures to form ammonia—or as by-products from coke ovens. Despite the increasing dominance of atmospheric nitrogen as a source of nitrogenous compounds, a small but economically healthy production of natural nitrates from the extensive Chilean deposits continues. Although natural nitrates are not essential mineral resources, the synthetics being more than adequate as replacements when necessary, their occurrence as *nonmarine evaporites* is both interesting and unusual.

Almost one-third of the nitrogen consumed annually is produced in the United States—9,100,000 tons compared with a world total of 31,700,000 in 1967. The Chilean production of 887,000 tons is small by comparison.

Sulfur Fertilizers

Because of its diverse uses, sulfur is not commonly thought of as a fertilizer element; approximately 40 per cent of the world's production, however, is used in the manufacture of superphosphate and ammonium sulfate, both essential fertilizers. The second-largest consumer, the chemical industry, takes 20 per cent and uses much of this for the production of insecticides and fungicides for crop protection.

Since sulfur is an abundant element, its sources are varied and widespread. The sea, for example, contains vast resources of *sulfates,* and the huge marine evaporite deposits of the world, discussed previously, contain enormous resources of $CaSO_4$. However, sulfur, like other resources, is preferentially sought in its least costly supply, this being the *native* form.

There are only two important resources of native sulfur. One, exploited principally in Japan, is from volcanoes, which give off sulfurous gases that condense in near-surface veins and rock impregnations. The other source, quantitatively much larger, is derived by secondary concentration from $CaSO_4$. Certain anaerobic bacteria derive their oxygen from solid compounds, such as $CaSO_4$, and their food supplies from decayed organic matter. A series of reactions ensues, but it can be summed up by the equation:

$$CaSO_4 + 2C + H_2O \xrightarrow{\text{bacteria}} CaCO_3 + H_2S + CO_2$$

$$\underset{\substack{\text{(for} \\ \text{energy)}}}{} \underset{\substack{\text{(for} \\ \text{food)}}}{}$$

When the H_2S comes into contact with oxygen, it is immediately oxidized to water and sulfur by the reaction:

$$H_2S + O \longrightarrow H_2O + S$$

The places where this process occurs are those where petroleum, to supply the bacterial food, is in contact with $CaSO_4$ deposits. In the United States and Mexico this has occurred in a strange environment: on top of salt domes. The origin of salt domes will be discussed later; for the present it is sufficient to state that the source of the salt is deeply buried marine evaporites, and that the associated $CaSO_4$ is brought to a near-surface environment where the bacterial reduction proceeds. Although several hundred salt domes have been found around the world, only a few contain commercial quantities of native sulfur. These occur along the coast of the Gulf of Mexico from Alabama to Mexico; they are very rich, however, and account for 25 per cent of the world's current production. The same bacterial reduction process occurred in petroliferous marine basins in Sicily, where the resulting sulfur deposits have been worked by hand for many centuries.

The distribution and size of the native sulfur reserves are not large, and since World War II a determined drive on the part of many countries has opened two alternative sulfur sources. The first is the H_2S, or *sour gas*, content of natural gases being recovered for fuel. Formerly allowed to escape, the H_2S component is now widely recovered and oxidized to sulfur. The second source is in the sulfide ores of the scarce metals, recovered as a by-product, and in deposits of two iron sulfides, pyrite (FeS_2) and pyrrhotite (FeS). Production from sulfide ores, principally by the U.S.S.R., Japan, Spain, and China, reached 9,700,000 tons in 1966.

The world production of sulfur from all sources reached 27,011,000 tons in 1967; of this, 9,316,000 tons were produced by the United States. Reserves of rich native sulfur ores are small, but resources of sulfide and sulfate ores are huge. Provided that technological development can keep the price of sulfur low as the alternative resources are used, we shall have abundant supplies.

Minerals for Chemical Supplies

The nonmetallic minerals used principally for raw materials in the chemical industry are diverse and have considerable economic importance. They have few claims to being essential compounds, however, and the use of most is guided chiefly by their great abundance, easy recovery, and hence low cost. Substitutes are already available for many.

The most important member of the group is salt (NaCl), known to geologists as the mineral halite. Not only does the sea contain vast resources of salt (see Chapter 2), but the marine evaporites also contain such vast deposits that the problems of man's resources essentially become problems of the practicality of mining and shipping (Fig. 5–5). The previously mentioned Permian evaporite basin of the South and Central United States, for example, contains more than 100,000 square miles of halite beds aggregating 200 feet in thickness, but the present cost of mining the deposits is high because of their depth below the surface as well as their distance from major markets; they are exploited only in one small area in Kansas.

Although many beds are too deep to mine, nature has an interesting way of bringing some salt nearer the surface. Halite has a density of 2.2 grams per

FIGURE 5-5 *Areas of the United States and Canada known or believed to be underlain by major marine evaporite deposits of salt. (After U.S. Geol. Surv. Bull. 10195, and A. D. Huffman, 1968.)*

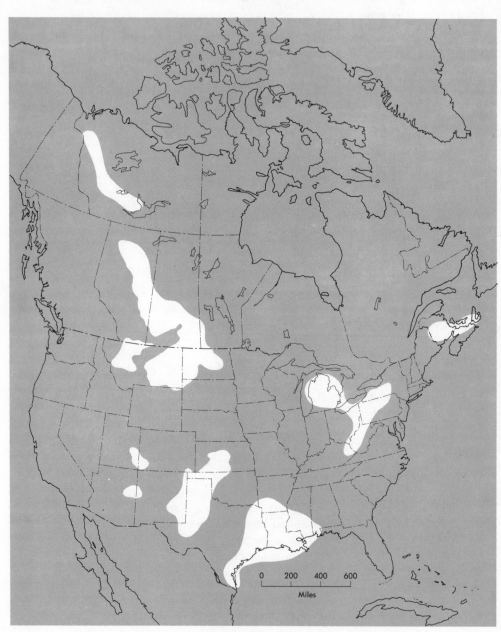

Industrial minerals: chemical and fertilizer supplies

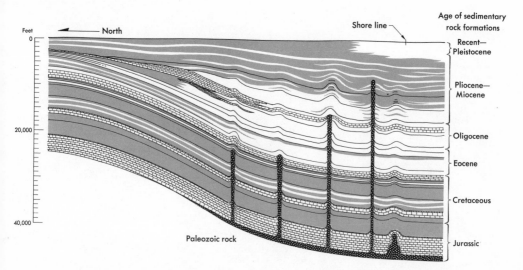

FIGURE 5-6 *Salt domes, narrow plug-like columns of salt that are believed to rise and "flow" upward through heavier but mechanically weak overlying rocks, bring deeply buried salt from marine evaporite deposits to near-surface environments. This geologic section of eastern Louisiana shows known salt domes that are believed to have risen through as much as 40,000 feet of overlying sediments. (After J. Ben Carsey, 1950.)*

cubic centimeter, whereas most of the associated sedimentary rocks have densities of at least 2.5 grams. The salt beds, being lighter and capable of plastic flow like ice in a glacier, tend to rise and "flow" up through the overlying rocks. Provided that the overlying rocks are still weak enough to be ruptured by the rising salt, long thin columns or plugs of salt will float their way up from deeply buried sedimentary salt horizons. Columns ranging from a few hundred feet to more than a mile in diameter (Fig. 5-6) are known to have risen up through as much as eight miles of overlying sediments.

Salt domes are known in many areas of the world—Europe, South America, the Mid-East, and the U.S.S.R.—but they are particularly frequent in the area bordering the Gulf of Mexico, where several hundred have been identified (Fig. 5-7). These domes are believed to have risen from evaporite beds 40,000 feet below the present flat coastal plain of Louisiana, Texas, and Mexico, and several are mined for salt in Louisiana and Texas.

Salt is an essential ingredient in our diets, and 99 countries produce it on a regular basis. Direct human consumption is only a small portion of the total, however, since most goes for the manufacture of chlorine and soda ash (Na_2CO_3) in the chemical industry, and for road control of ice and snow. Not surprisingly, the major industrial powers—the United States, the U.S.S.R., Germany, and the United Kingdom—consumed 50 per cent of the 1966 production of 123,000,000 tons.

Of the raw chemical group, with the exclusion of petrochemicals—which are derived from fossil fuels—salt is the most largely produced member. Others

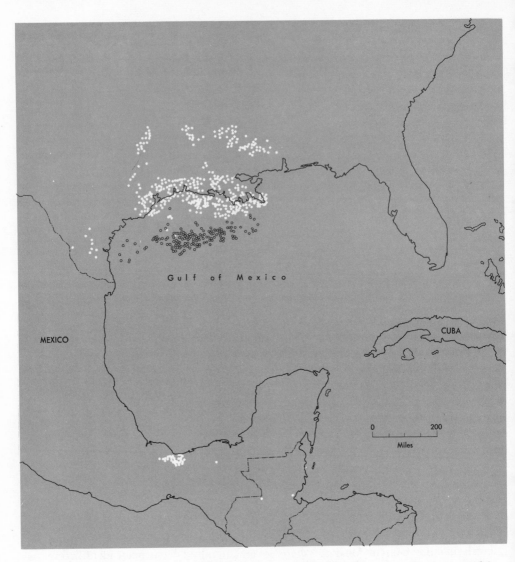

FIGURE 5–7 *Known and probable salt domes (white dots and open circles, respectively) that have been identified in the coastal area of the Gulf of Mexico. In no other part of the world have so many of these unusual bodies been identified. (Adaptation of Fig. 5.1, p. 203,* Geology of the Atlantic and Gulf Coastal Province of North America *by Grover E. Murray [Harper & Row, 1961].)*

of importance are Na_2CO_3, used for production of paper, soap, and detergents, and for water treatment; Na_2SO_4, used for kraft paper, detergents, and additives in tanning and dyeing; and borate minerals, such as borax ($Na_2B_4O_7 \cdot 10$ H_2O), used for glass-making fluxes, soaps, detergents, hide-curing compounds, and antiseptics. These deposits are all formed by precipitation from lake waters in *nonmarine evaporite deposits* and are available in large quantities.

Industrial minerals: chemical and fertilizer supplies

Other Industrial Minerals

The list is large but only one group, the abrasive minerals, is so technologically essential that it requires discussion.

Abrasives are becoming increasingly more important for working the hard and closely machined modern alloys and compounds such as tungsten carbide, and it is diamond, the hardest known natural compound, that is the most important resource. Diamond—the most dense natural form of carbon—requires high pressures for its formation; these are reached only at depths of 100 miles or more in the Earth. The diamond-bearing rocks from these great depths, called kimberlites, are very mafic and are themselves rare. They reach the surface in narrow pipe-like vents, often no more than 100 feet in diameter, and the reasons for the formation and location of the pipes remain a geological mystery.

Kimberlites are the home of the diamond, but not all kimberlites contain diamonds. Of the several hundred kimberlite pipes found in Africa to date, only 27 are known to contain diamonds, and of these, 11 are too lean to warrant mining. On other continents the percentage is even lower, and, outside Africa, only in the Yakutia area of the U.S.S.R. have kimberlites been discovered that are rich enough to be worked. Even the richest kimberlites contain very low diamond contents—no more than 1 carat[1] for every 3 tons of rock, which is equivalent to 0.0000073 per cent.

Because diamonds are dense and almost indestructible, they accumulate in placer deposits; 96 per cent by weight of all diamonds recovered still comes from placers (Fig. 5-8). Though we immediately think of gems when diamonds are mentioned, only 20 per cent of the diamonds produced can be so cut—though this still amounts to 65 per cent of the monetary value of diamond production—and the remaining 80 per cent is used for such industrial purposes as cutting, die-making, and the manufacture of abrasives.

From the time of their discovery in 1870, African deposits have been the world's largest producers of diamonds—first in South Africa, with kimberlite pipe production unusually rich in gem material, but more recently in the Congo and in Ghana—and these have produced over 80 per cent of the world's diamonds (Table 5-4). Recent finds in northern Siberia account for a currently growing Soviet production. Measured reserves of diamond, largely in the Congo, are reported to be in excess of 600,000,000 carats.

As it has with other abrasives, technological progress has finally shown the way to self-sufficiency in diamonds. In 1955 the General Electric Company

[1] The carat is a unit of weight used for precious stones; 1 carat = 200 mg. or 0.0070 oz.

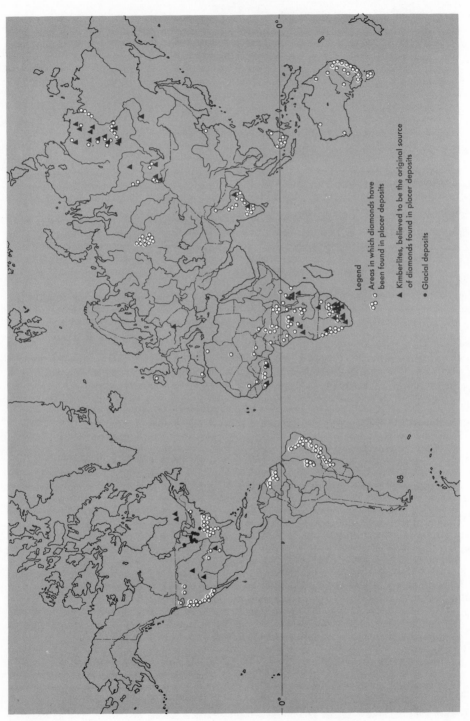

FIGURE 5-8   Areas of the world in which diamonds have been found. Only the African, Siberian, and South American areas have recorded significant productions in recent years. (After D. P. Gold, 1968.)

Legend

○⊖ Areas in which diamonds have
   been found in placer deposits

▲ Kimberlites, believed to be the original source
   of diamonds found in placer deposits

● Glacial deposits

Table 5–4

Major Industrial Diamond-producing Countries, 1967

Country	Production (Carats)	Percentage of Total
Congo	17,890,000	53.7
U.S.S.R.	5,600,000	16.8
Republic of S. Africa	4,900,000	14.7
Ghana	2,283,000	6.9
World total	33,295,000	

U.S. Bureau of Mines.

announced its successful synthesis of industrial diamonds, using a special ultra-high-pressure reaction vessel. Commercial production has grown to a presently estimated level of 10,000,000 carats per year, with production coming from Sweden, South Africa, Ireland, Japan, and the Soviet Union, in addition to the United States.

6

Industrial rocks:

the building materials

There will be a shortage of standing room on earth before there is a shortage of granite. (J.A.S. Adams, New Ways of Finding Minerals, *1959.)*

Building materials are the largest tonnage crop and, after fuels, the second most valuable commodity that we reap from the Earth. Because almost every known rock type and mineral contributes to the crop in some way, the origin of building materials embraces most of geology. We will concentrate mainly on uses and try to place this largest-volume mineral production into perspective with the other mineral products.

Although no classification is completely satisfactory, we will separate building materials into two groups. First, materials that are used as they come from the ground, without any treatment beyond physical shaping, such as cutting or crushing. Second, the prepared materials that must be treated chemically, fired, melted, or otherwise altered before use so that they can be molded and set into new forms. The first group includes building stones, sand, gravel, and crushed stone for aggregate; the second includes clay for bricks, raw materials for cement, plaster, and asbestos.

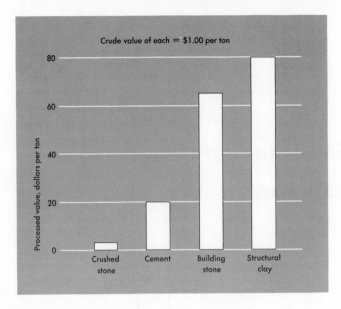

FIGURE 6-1 *Processing of building materials adds greatly to their value. (After W. L. Fisher, 1965.)*

The building materials have little value in the ground; they are not scarce commodities and are widely distributed, but when removed and processed to a useful form, they increase in value enormously. For example, the limestone and shale used to make cement may have intrinsic values of only $1 per ton or less in the ground, but after mining, crushing, firing, and conversion to a high-quality cement, the product is worth $20 or more per ton (Fig. 6-1). The factors controlling location of production sites are most commonly such straightforward economic problems as local demand; rarely are they problems of resources.

Natural Rock Products

There are three important classes of natural rock products, and the supplies of each are almost limitless. It makes little sense to consider global reserves, therefore, since production problems derive from costs of transportation rather than from abundance.

Building Stone

The uses of building stones range from roofing slate and curbstones to facings for public buildings and tombstones. Though used from time immemorial as structural and foundation materials, as in the Egyptian pyramids, building stone is being supplanted by concrete; its remaining uses largely are governed by the pleasing ornamental display of natural stone.

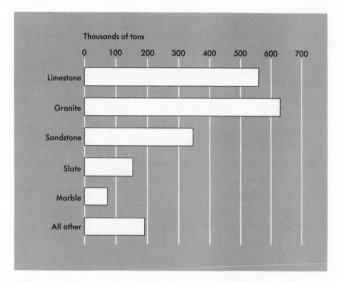

Thousands of tons

0 100 200 300 400 500 600 700

Limestone
Granite
Sandstone
Slate
Marble
All other

FIGURE 6–2 *Types of building stones used in the United States in 1967. Although direct figures are unavailable, it is probable that the same approximate range of rock types is used throughout the world. (After U.S. Bureau of Mines.)*

Figures on production volume and types of stone used are not available on a world-wide basis; however, the best-documented production—that from the United States—probably gives a reasonable idea of the global balance. The total tonnage of building stones mined in the United States in 1967 was 2,011,000 tons. The two most popular rock types were limestone and granite, jointly aggregating more than 60 per cent of the total (Fig. 6–2).

What are the problems with building stones? When stones of pleasing appearance and good physical properties have been found, problems mostly are to do with mining: how to mine the rock without shattering it—which means that extensive blasting is not possible—and how to select areas where natural joints and cracks in rocks are a help, rather than a hindrance, to mining. The answers to these and related problems require solutions that may vary from the most primitive to the most advanced (Figs. 6–3A, B, and 6–4).

Crushed Rock

A commodity of enormous proportions, crushed rock accounted for 811,-000,000 tons in 1966 in the United States alone. Yet little more than a century has passed since Eli Whitney Blake, responding to the call for quantities of crushed rock needed for the then ambitious project of a two-mile-long macadam road from New Haven to Westville, Connecticut, invented the modern rockcrusher in 1858. Prior to this, all rockcrushing was done by hand.

Crushed stone is still used principally for roadbeds and for concrete aggregate, although 91,456,000 tons, principally of limestone, were used as raw ma-

A

Figure 6–3 (A) A present-day granite quarry near Madras, India, where the earliest quarrying methods are still in use. The initial rock breakage is accomplished by slowly heating the open rock face. Thermal expansion causes the outer layers to expand differentially and eventually to break free from the underlying rock. The depth of the break can be controlled by the size of the fire and consequently by the rate of heating. (B) The spalled section is worked into building-stone units by hammer and chisel. (Photographs by author.)

B

FIGURE 6–4 *A modern granite quarry, in Barre, Vermont. The striated surface came from vertical drill holes two inches in diameter, drilled side-by-side to a described depth, and a small one-inch section between them drilled out to make a continuous cut. The horizontal cut is made by drilling a series of holes a foot apart, then firing a small charge of gunpowder to open a break while avoiding fracture of the mass. (Courtesy Rock of Ages Corp.)*

terial for the manufacture of cement in 1967. The most widely used rock types are limestone and dolomite, both easy to mine and crush, but strong in use. Basalt and other fine-grained, dark-colored igneous rocks, commonly called trap rock by industry, came a poor second (Fig. 6–5). The only essential requirement for crushed stone is a rock outcrop to quarry, and indeed rock-crushers can be seen—and heard—adjacent to most cities in developed countries.

Sand and Gravel

Sand and gravel, largely used like crushed rock for highway roadbeds and for concrete aggregate, are consumed in amounts even larger than those of crushed stone. The consumption in the United States in 1967 was 905,162,000

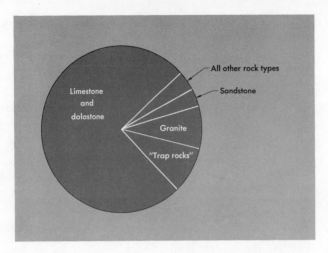

FIGURE 6–5 *Common rock types most widely used in the preparation of crushed stone in the United States. (After U.S. Bureau of Mines.)*

tons. Sand is classified as having particle sizes less than 0.079 inches or 2 mm in diameter; those of gravel are larger. The tonnage of gravel used is twice that of sand.

The geological domain of gravel is that of streams, where the rounded pebbles are produced by continuous movement of fast water, and where a size separation occurs on the basis of weight: finer grains are washed downstream or out to sea. Local resources of sand and gravel may often be limited, especially in flat regions devoid of large rivers. Where land resources are being depleted too rapidly, as in New York City, exploration and exploitation of off-shore marine sandbars may also be employed.

Prepared Rock Products

From the day man first molded a clay object and fired it in his hearth, he has been using—and continuously expanding his use of—prepared rock products. Although we now process and use a bewilderingly large array of materials, four of these account for most of the volume and value.

Cement

The word cement refers to agents that bind particles together. Cement is rarely used alone; it is added to sand, gravel, crushed rock, or other aggregate as the binder needed to make concrete, a sort of "instant rock." As little as 15 per cent of a concrete may be cement.

The forerunner of modern cement was discovered by the engineers of ancient Rome. They found that water added to a mixture of quicklime (CaO, obtained by heating—or calcining—limestone) and a natural, glassy volcanic ash from the town of Pozzuoli, near Naples, produced a series of reactions that caused the mass to recrystallize and harden. The resulting mass was stable in air or water and materially assisted the Romans in their remarkable engineering feats. Long known as *pozzolan cement,* similar materials are still employed today, but on a decreasing scale. In a pozzolan cement one of the ingredients (quicklime in ancient Rome) is calcined to get it in a reactive form; the others are naturally reactive materials. However, the volcanic ash had also been essentially calcined by nature during the volcanic eruption; why should all the ingredients not be calcined by man if the right rock compositions were selected? Apparently the Romans realized this too, for they discovered the secret of cement manufacture.

The "secret" was forgotten during the Dark Ages, however, not to be rediscovered until 1756. John Smeaton, a British engineer engaged in designing and building the famous Eddystone Lighthouse, sought cementing materials to set and remain stable under water. He is said to have rediscovered the Roman hydraulic cement formula through examination of an ancient Latin document. Natural cements soon became popular in Europe and subject to much experiment. In 1824 another Englishman, Joseph Aspidin, patented his formula for *portland cement,* so called because it resembled Portland stone, a limestone widely used in British buildings. It soon supplanted all other cements and is today the most common construction material in the world, with twice as much concrete being used as all other structural materials combined.

Production of portland cement has grown erratically, but continuously, at a high rate (Fig. 6–6) as population and technological expansion both increase; and the rate is expected to grow. Not surprisingly, the largest producers and consumers are the highly industrialized countries (Table 6–1).

FIGURE 6–6 *Growth of the portland cement industry in the United States as shown by the continuously increasing amounts consumed in this century. Maximum growth occurred in periods of greatest industrial and transportational construction activity. (After U.S. Bureau of Mines.)*

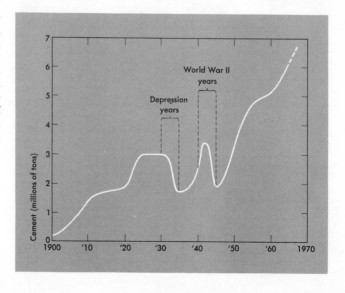

Portland cements, which have a range of composition depending on their use, are prepared by heating finely ground rock of suitable composition to a temperature of approximately 1,480°C, when all the carbon dioxide and water are expelled and part of the charge melted to a glass. The resulting clinker is again crushed and is then ready for use; when it is mixed with water, a series of rapid chemical reactions proceed in which new compounds form and grow as a hard cemented mass of interlocking crystals.

The raw materials for portland cement (Fig. 6–7) are found in a suitable mixture of a slightly dolomitic limestone and a shale or clay. As limestone is the largest ingredient, cement works are usually situated close to a suitable source and other ingredients are transported in. The most desirable circumstance is a somewhat impure limestone in which the impurities are clays of the desired composition. Such beds are indeed found, and natural cement rocks now account for approximately 20 percent of the cement production.

Table 6–1

Seven Largest Producers
of Portland Cement, 1967*

Country	Millions of Tons	Percentage of Total
U.S.S.R.	93.5	17.7
United States	72.5	13.7
Japan	47.4	8.9
West Germany	34.7	6.6
Italy	29.0	5.5
France	27.1	5.1
United Kingdom	19.4	3.7
World total	529.0	

U.S. Bureau of Mines.
* 104 different countries produced portland cement, but these seven large industrial nations accounted for 61 per cent of the production.

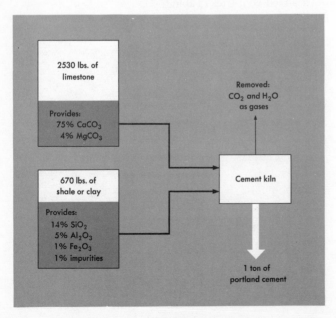

FIGURE 6–7 *Typical feed of raw materials needed to produce a widely used composition of portland cement.*

Plaster

Plaster, made by heating or calcining gypsum ($CaSO_4 \cdot 2H_2O$), is one of man's oldest building materials, and the consumption continues to grow, with a world-wide production of 49,629,000 tons in 1967. Calcining gypsum at 177°C rapidly drives off 75 per cent of the water and changes it to a new compound, $2CaSO_4 \cdot H_2O$, commonly called *plaster of Paris*, after the famous gypsum quarries in the northern part of that city, from which a plaster of particularly high quality is produced. When plaster of Paris is mixed with water, it reverts to a finely interlocked mass of tiny gypsum crystals. Plaster can be used straight, mixed with sand or fine-grained aggregate, or mounted on wallboard or paper backings for prefabricated finished surfaces.

In Chapter 5 we discussed the precipitation of $CaSO_4$ in evaporite sequences. The precipitate may take one of two forms, either gypsum or anhydrite ($CaSO_4$), so named because it does not contain any water of crystallization. Which of the two forms precipitates is a function of temperature, with anhydrite being favored by higher temperatures; under the hot climatic conditions where most evaporites form, anhydrite is the common precipitating form. It is gypsum that we usually find today, however, for the ground temperature in the shallow deposits where it is worked is low, and with water percolating down from rainfall, the anhydrite soon hydrates to gypsum in the same way that plaster of Paris does.

Gypsum and anhydrite are widely distributed; more common than salt but less common than limestone, they are known throughout the geologic record. A resource map published by the U.S. Geological Survey shows that as much as 10 per cent of the land area in the United States is underlain by gypsiferous rocks, which indicates that potential resources are so large as to be superabundant. Indeed, the U.S. Bureau of Mines has stated that measured reserves in the United States alone are sufficient for 2,000 years at present rates of production.

Clays

There are so many structural and refractory ceramic materials now used that discussion of each would be pointless. Most are formed from clay that can be molded into desired shapes, then fired to hardness. Bricks so formed have been used at least since the days of the Babylonian Empire, first in a form with low durability—the bricks having been allowed to dry in the sun—but later in forms of increasing durability, achieved by firing and even by glaze-coating. As with cements, the art of brickmaking and clay work reached a high degree of perfection in the days of the Roman Empire, but much of the art was lost in Europe during the Dark Ages. From the time of the industry's revival in Europe in the thirteenth century, production of clay for structural ceramics has grown steadily—first for the preparation of bricks, but later for tiles, drainpipes, and numerous other uses. By 1967, the production of clay in the United States alone was 54,664,000 tons, and of this, 57 per cent was for structural ceramics, mostly bricks, and only 34 per cent for production of cement.

Clays are formed by weathering at the Earth's surface and may accumulate as residual deposits, as has been discussed in "Aluminum" (Chapter 3); they also may be transported and deposited as sedimentary clay beds. Whether transported or residual, clays soon begin to lithify and become solid rocks when they are dehydrated and heated by burial. Their environment is the Earth's surface and not the deeper portions of the crust. Like the other building resource materials, therefore, they are usually recovered by large quarrying operations in surficial deposits, with the quarries preferentially situated as close to major consuming sites as possible.

Known and exploited resources of clays are so large that few countries make realistic attempts to estimate them. Those of the United States, for example, are said by the U.S. Bureau of Mines to be more than sufficient for another century of production at anticipated rates of growth.

Glass

We might not immediately think of glass as belonging in a category with bricks and cement, but it consumes similar raw materials and is processed in a similar fashion. Furthermore, glass has begun to challenge older, more established structural materials in many specialized uses, and consumption is rising rapidly.

Glass is made by melting rocks and minerals, then quenching them so rapidly that crystals do not have time to nucleate. This procedure is more readily carried out with some materials than with others, and most readily with silica (SiO_2), usually obtained from quartz in sandstones. The melting point of quartz is very high ($1,713°C$), and to reduce this to a more easily reached temperature, ingredients such as CaO (from limestone), Na_2O (from sodium carbonate), and borax are added.

Asbestos

Asbestos is the name given to the fibrous forms of many minerals. One such, chrysotile ($Mg_3Si_2O_5(OH)_4$), a fibrous form of the serpentine minerals, is so abundant, and has such excellent physical properties, that it accounts for more than 90 per cent of all the asbestos minerals mined. Asbestos fibers, being

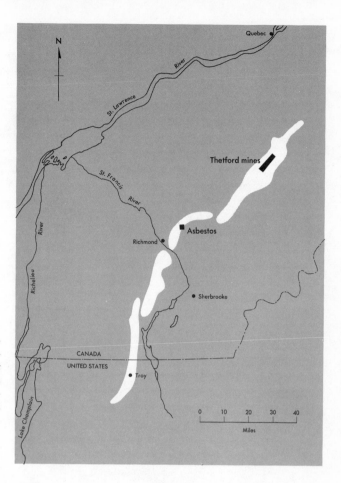

FIGURE 6–8 *Belt of serpentinized peridotites crossing Quebec and Vermont from which most of the North American asbestos production comes. (After S. B. Keith and G. W. Bain, 1932, Economic Geology, v. 27, p. 69, with permission of the publisher.)*

strong and flexible, can be spun and woven like organic fibers such as cotton and wool. The resulting products are not flammable, are good electrical and thermal insulators, have excellent wear-resistant properties, and are stable in many corrosive environments. Fibers more than three-eighths of an inch in length are preferred for thread-making, and it is these that are used for electrical insulation and other specialized uses. The large volume of short fibers produced is bound by inert media, such as portland cement (to produce roof and wall shingles) and vinyl plastics (to produce tiles).

Chrysotile asbestos apparently forms large deposits in only one way: by the near-surface hydration and alteration of peridotite,—an ultramafic igneous rock rich in the magnesian olivine, forsterite (Mg_2SiO_4)—to form serpentine. The world's largest producer, Canada, obtains its supplies principally from a belt of serpentine running northeasterly across Quebec, and extending into Vermont where there is small-scale production (Fig. 6–8). Although serpentines are not uncommon rocks, asbestos resembles the scarce metals in its supplies, for only a tiny fraction of the serpentines have large, commercially exploitable

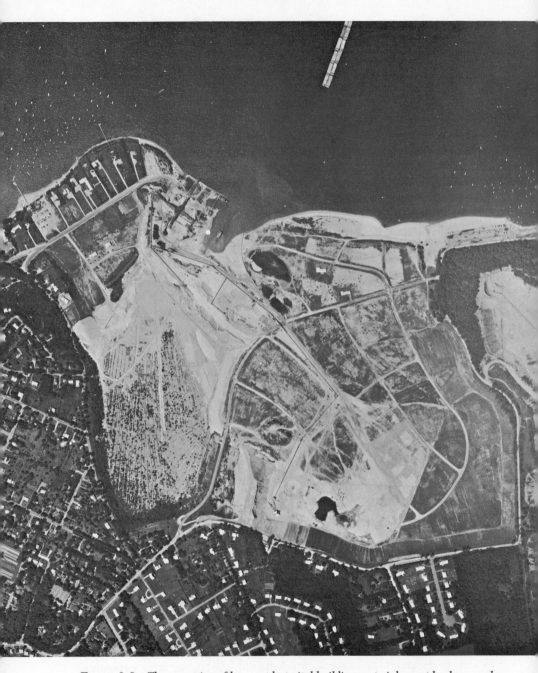

FIGURE 6-9 *The quarrying of low-cost but vital building materials must be done as close as is practicable to the consuming city. The city may later grow out to and around the older quarry site, thus becoming a possible cause of irritation between producer and consumer. The Steers Sand and Gravel Corp., New York, N. Y., for example, now finds itself bounded by suburban development and recreation areas. (Courtesy National Sand and Gravel Association.)*

Industrial rocks: the building materials

deposits. The world's largest reserves of asbestos are in the hands of the four largest producing countries — Canada, the U.S.S.R., Republic of South Africa, and Southern Rhodesia (Table 6–2).

Asbestos is one of the resources in which the United States is poorly endowed; although it consumes approximately 25 per cent of the world production, it produces less than 4 per cent, and must import its remaining needs from Canadian and African sources.

Building Supplies and City Growth

Table 6–2

Leading Asbestos Producers of the World, 1966*

Country*	Production (Tons)
Canada	1,479,000
U.S.S.R.	925,000
Republic of So. Africa	276,000
Southern Rhodesia	175,000
World total	3,350,000

U.S. Bureau of Mines.
* These four countries also contain the largest known asbestos reserves.

It is unfortunate that exploitation of mineral resources produces scars on the Earth's surface that are not easily eradicated; these scars have commonly been left for nature to heal, but in a populous world this is no longer always possible. The costs of rehabilitating a mined-out area are rarely reckoned among the costs of recovery, and because it often falls to governments to clean up after mining operations have ceased, an additional unspecified cost is paid out of the public purse. Although this problem is common to all mining industries, it is often most apparent in the production of building supplies.

The annual volume and tonnage of building supplies needed are, as we have seen, enormous. The supplies are widely available, and it is wise to reduce transportation costs by locating supply quarries as close as possible to the consuming cities, the centers of population growth; most of these have grown around and over large, unsightly quarries that were once situated in an inconspicuous place on the cities' peripheries (Fig. 6–9). The conflicts that develop are all too common. The problem cannot be solved simply by pushing the producing centers far from consumption sites, for transportation costs soon become prohibitively high. Prudent city and urban planning, recognizing the need for inexpensive resources, and perhaps requiring a small tax to facilitate future rehabilitation, will become increasingly pressing as cities expand.

7

Energy:
the saga of the fossil fuels

We in the United States have been . . . fabulously successful in discovering and developing . . . multipliers and additives for our supply of human energy. Just 100 years ago 94% of the power used in industry had to come from the backs of men—whereas today, human power accounts for something less than 8%. (Thomas J. Holme, "Produce and Compete or Perish," American Scientist, vol. 56, 138–158, 1968.)

The energy flux at the Earth's surface is large and is channelled in many different paths (Fig. 7–1). Man naturally occupies a small place on the path that begins with solar energy captured by plants, but he is unique among animals in his ability to utilize energy from other paths to increase the effectiveness of his labors. By domesticating herbivorous animals and by burning wood, he has long drawn supplemental energy from the same bank as his own food energy, the living plants. Increasingly great usage is now being made of other supplemental energy sources, however, and three of these—the buried plant remains (commonly called fossil fuels), the water reservoirs, and the nuclear fission bank—have become vital resources for the continuity of our complex civilization.

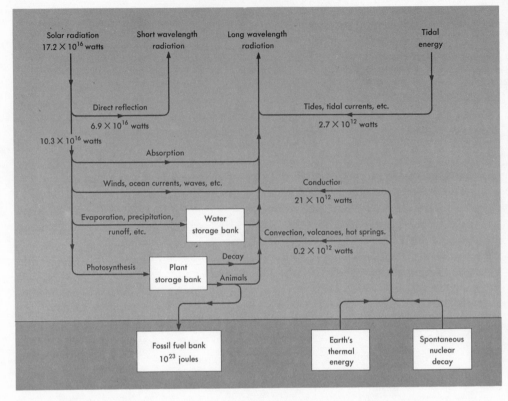

FIGURE 7-1 *An energy flow sheet for the Earth. The unit of energy used is the joule. Where the energy flux is continuous, such as that flowing in from the Sun, the absolute power unit, the watt, is used to measure the flux. A watt is equal to one joule per second. All the energy temporarily stored in the fossil fuel bank, therefore, is equal only to the energy radiated to the Earth every seven days by the Sun. (After M. K. Hubbert, Energy Resources, Pub. 1000-D, Committee on Natural Resources, Nat. Academy of Sciences—Nat. Research Council, Washington, D.C. 1962.)*

The increasingly wide availability of supplemental energy sources has led to a dramatic and continuously growing energy consumption. In the United States, for example, the total energy consumption rose from 8×10^{15} Btu[1] in 1900 to 50×10^{15} Btu in 1965 (Fig. 7–2). The sources of our energy supplies have changed dramatically during the last century. Whereas 70 per cent of the energy used came from wood in the plant storage bank in 1870, we draw less than 0.5 per cent from the same source today (Fig. 7–3). The prime energy sources for the present and immediate future are fossil fuels—coal, oil, and

---

[1] In comparing energy sources it is essential to use the same energy units. Because thermal energy from coal, oil, and gas are so important, we will use the British Thermal Unit (Btu), which is defined as the heat energy needed to raise the temperature of 1 pound of water by $1°F$, and the gram calorie, which is the heat energy needed to raise the temperature of 1 gram of water by $1°C$. 1 Btu = 252 gram calories.

The absolute unit of work is the joule, equal to 0.2390 gram calories or 0.000948 Btu. The absolute unit of power is the watt, equal to the consumption of 1 joule of energy per second.

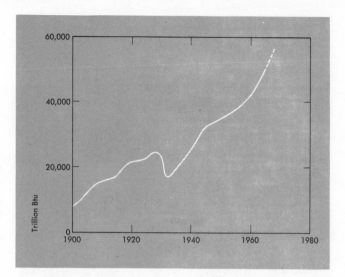

FIGURE 7-2 *Growth of energy. Production in the United States from mineral fuels and water power. During the 65-year period when power consumption grew eightfold, the population increased approximately two and a half times. (After U. S. Bureau of Mines.)*

natural gas—and, to a much smaller extent, water power. Fossil fuels are non-renewable, and their depletion is already cause for concern to many long-term economic planners. In response to the concern, the nuclear fission energy bank —which, although also nonrenewable, is much larger than the fossil fuel bank— is widely accepted as a great hope for future energy supplies. Other sources, such as solar, tidal, and geothermal power, are used where convenient, but unfortunately they lack the flexibility in use and transmission that fossil and nuclear fuels enjoy, and they are not generally considered to be of more than local importance for the foreseeable future.

Fossil Fuels

There are several forms of fossil plant remains, but only three are now of major importance as fuels: coal, crude oil, and natural gas; together they account for more than 95 per cent of all the energy generated in the United States (Fig. 7-3) and for an even higher percentage in most other countries.

The amount of solar energy temporarily stored in living plants by photosynthesis is very large, but the rate of decay in the Earth's oxidizing atmosphere and the consequent release of stored energy are nearly equal to the contemporary rate of photosynthesis. The energy is soon radiated back into space and lost. In a few favored places, such as swamps and bogs, plant matter is preserved in reducing environments that retard the decay process and lead to storage of a small fraction of the photosynthesized energy. The fraction preserved in any one growing cycle is tiny, but the accumulated organic remains (over the 600 million years that life has been prolifically distributed across the Earth) are now considerable.

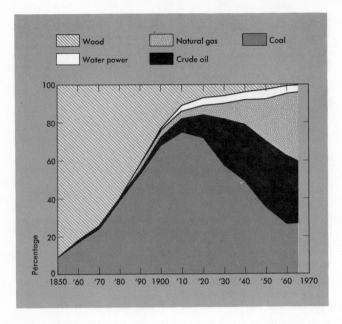

FIGURE 7–3 *Percentage production of energy from different fuels and water power in the United States. Nuclear power production is still too small to show on this plot. (After U.S. Bureau of Mines and S. H. Schurr, B. C. Netschert, 1960.)*

Coal

Coal is formed from the remains of fresh-water plants. Its apparent environment of formation is a densely vegetated, water-covered swamp. Into the stagnant swamps fall dead limbs, trunks, leaves, and spores, soon to be waterlogged and to sink. Once water-covered, they are protected from the atmosphere; bacterial digestion may continue, turning the woody plant remains to a jelly-like, amorphous mass of peat, but oxygen supplies are quickly consumed and decay ceases. Thick accumulations of peat can form only if the swamp basin slowly subsides, and rich deposits can occur only if the inflow of clays and other inorganic detritus, commonly called *ash*, is low. Suitable coal-accumulation sites are not widespread at present, but an apparently typical one does occur in the Dismal Swamp of Virginia and North Carolina, where an average of 6 feet of modern peat covers a 2,200 square mile area.

Coal reserves are widely distributed in the geologic record, however. The earliest fossil record of undoubted land plants—which proliferated rapidly after emergence—is in Upper-Silurian rocks, and they were widespread by Mid-Devonian times.[2] The earliest coal measures of significant size are found in Upper-Devonian rocks of the Canadian Arctic regions. In the two succeeding geological epochs, the Carboniferous and the Permian, occurred the most important coal-forming period in the Earth's history. Coal measures were formed on all continents, and the great deposits of eastern North America and Europe were laid down.

[2] The interested reader should consult the companion volume *The History of Life*, by A. L. Mc-Alester, for an expanded discussion of plant origins and distributions.

Peat, the first stage in the formation of coal, is a low-rank material, which is to say that it has a relatively low carbon content and a low heat-producing or calorific value. Upon burial and compaction of peat, a series of reactions occurs, and much of the water, oxygen, nitrogen, and other plant elements originally present is expelled, leaving an increasingly dense and carbon-rich coal. The process of coalification proceeds with age, bringing an increase in rank (Fig. 7–4), so that older coals generally have higher ranks than younger

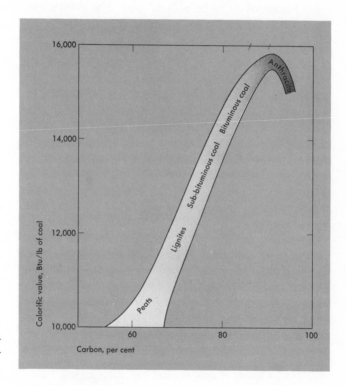

FIGURE 7–4 Increasing calorific value of coal with increasing rank.

ones. Naturally, the highest-rank bituminous coals and anthracite are the most highly prized fuels.

Coal measures are confined to sedimentary basins and in particular to those with fresh-water sediments. Geologists believe that all the major coal basins of the world are now discovered; on this basis, P. Averitt of the U.S. Geological Survey in 1969 estimated the world's inferred reserves of recoverable coal to be $8,415 \times 10^9$ tons. Averitt considered recoverable such coal as lay in seams greater than 12 inches thick, lying no deeper than 6,000 feet. He also assumed that the then most efficient mining recovery—50 per cent of the coal present—

was a reasonable rate. Despite these qualifications, the estimated tonnage is enormous, although the distribution of recoverable coal is highly erratic (Fig. 7-5), with North America, Europe, and Asia containing 97 per cent of the world's reserves (Fig. 7-6).

Coal was used on a small scale by the Chinese 2,000 years ago. It was also used by other isolated societies, such as the Hopi Indians who mined and burned an estimated 100,000 tons for fuel between the thirteenth and seventeenth centuries in the Jeddito Valley of Arizona; these uses did not spread, however. Use of coal in Europe as a widespread fuel began in the twelfth century A.D. when inhabitants of the northeast coast of England found that inflammable black rocks weathering out of coastal cliffs were good substitutes for their rapidly disappearing forest woods. Known as "sea coles," the new fuel soon became widely used—to the extent that by 1273 outraged Londoners complained of repugnant odors and air pollution arising from coal-burning! This warning deterred no one, however, and the use of coal as a fuel spread rapidly through Europe. Although coal as the predominant source of energy has now declined in oil-rich countries such as the United States (Fig. 7-3), it remains predominant throughout the world (Fig. 7-7).

Crude Oil and Natural Gas

Crude oil and natural gas, the liquid and gaseous components of petroleum, occur together, contain many of the same compounds, and have common origins concerning which there is still some controversy. The reasons of the controversy are not difficult to find: Petroleum is a derived material and has the characteristic of migrating from its site of formation; the precursors are therefore not always readily identified. With the recent rise of organic geochemistry—a new and specialized area of science—and with the development of new analytical tools for determining the organic compounds present in petroleum, the problem of origins comes closer to solution; however, it remains a prime area of research.

Crude oil and natural gas are composed chiefly of hydrocarbons[3] and, like coal, are found in sedimentary basins (Fig. 7-8). Although not confined to marine basins, petroleum is more abundant there than in fresh-water basins,

[3] Hydrocarbons are compounds containing carbon and hydrogen. There are many groups of hydrocarbon compounds, related by formula and structure, but the three of paramount importance in petroleum are: (1) The stable paraffin (alkane) series of saturated straight molecular-chain compounds, general formula C_nH_{2n+2}; methane (CH_4) is the commonest member. Saturated means a sufficiency of hydrogen atoms to satisfy the electron requirements of carbon atoms without the carbons sharing an extra electron to form a double bond. (2) The stable naphthene (cycloparaffin) series of saturated molecular-ring structures, general formula C_nH_{2n}; cyclopentane (C_5H_{10}) and cyclohexane (C_6H_{12}) are the most important compounds in petroleum. (3) The relatively unstable molecular-ring structures, general formula C_nH_{2n-6}; benzene (C_6H_6) is the most important petroleum compound.

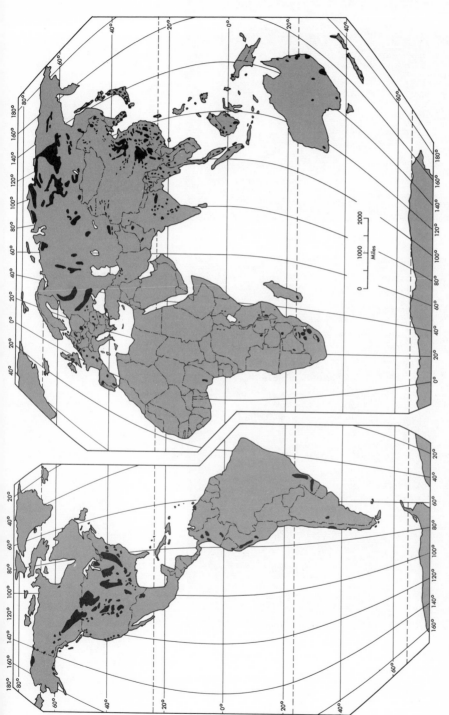

FIGURE 7-5 Occurrence of coal measures around the world has a very erratic distribution, with most known reserves occurring in North America, Europe, and Asia. (After Oxford Economic Atlas of the World.)

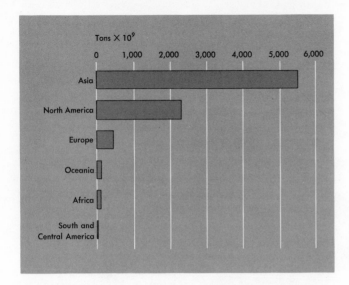

FIGURE 7–6 *Geographic distribution of recoverable coal reserves. (After P. Averitt, 1969.)*

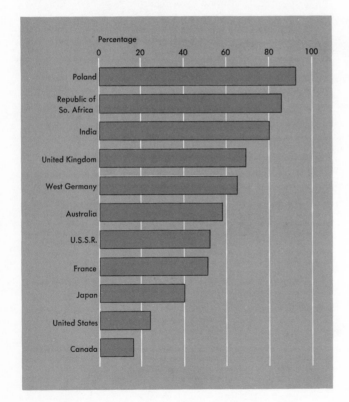

FIGURE 7–7 *Percentage of total energy supplied by coal in selected countries, 1964. (After H. Perry, 1967.)*

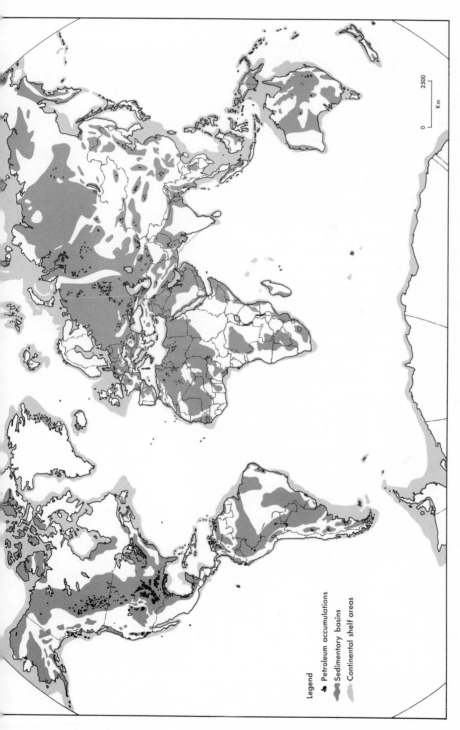

FIGURE 7–8   Distribution of the sedimentary basins of the world and the petroleum accumulations that have been located in them. The continental shelf areas—the sea-covered portion of the continental crust—are potential sites of future petroleum discoveries, particularly where they contain seaward projections of the sedimentary basins. (After International Petroleum Encyclopedia, 1968.)

Legend

- Petroleum accumulations
- Sedimentary basins
- Continental shelf areas

0   2500
    Km

and is conspicuously more abundant in basins with a high percentage of organic-rich sediments. Almost all sediments contain some organic debris, however, and a wide variety of petroleum hydrocarbons and even droplets of crude oil have now been found associated with more or less unaltered organic matter in many common rocks such as shales and limestones. There is little doubt, therefore, that widespread sedimentary organic matter, of microscopic plant and animal origin, is the source of petroleum; and further, it would seem that petroleum formation begins immediately after deposition of the organic matter.

The earliest-formed petroleum compounds tend to have high molecular weights, like those in the living cells from which they are derived, and produce "heavy" oils. On burial, and as the temperature and pressure rise, the large molecules are broken, or *cracked*, into lighter and more mobile ones. The longer the process continues, the "lighter" the crude oil becomes. Although the overall chemical composition does not change much, and most crude oils and natural gases fall in a small bulk chemical composition range (Table 7-1), the diversity of individual compounds produced is so great that no two oils ever contain the same molecular mix. Indeed, so complex are the mixtures that no natural crude oil has yet had its complete compound spectrum determined.

Table 7-1

Chemical Composition
of Typical Petroleum

Element	Crude Oil	Natural Gas
Carbon	82.2–87.1%	65–80%
Hydrogen	11.7–14.7	1–25
Sulfur	0.1– 5.5	trace–0.2
Nitrogen	0.1– 1.5	1–15
Oxygen	0.1– 4.5	—

From *Geology of Petroleum*, Second Edition, by A. I. Levorsen. W. H. Freeman and Company. Copyright © 1967.

The lighter and more mobile the petroleum hydrocarbons become, the more readily they apparently migrate. Though the exact mechanisms of migration remain uncertain, oil, gas, and water in sedimentary rocks move and slowly escape towards the surface; where barriers or traps are interposed in the migration paths, accumulations result, (Fig. 7-9). The fact of slow escape is substantiated by the observation of Weeks and others that the highest ratio

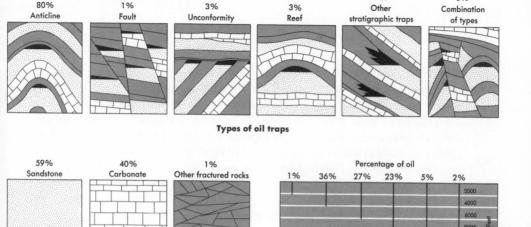

Types of oil traps

| 80% Anticline | 1% Fault | 3% Unconformity | 3% Reef | 7% Other stratigraphic traps | 6% Combination of types |

Type of reservoir rock

| 59% Sandstone | 40% Carbonate | 1% Other fractured rocks |

Depth of oil trap

FIGURE 7–9 *Types of oil traps and reservoir rocks, and the depth of known oil pools, together with the percentage of the world's oil production from each.* (From Man's Physical World by J. E. Van Riper. Copyright © 1962 by McGraw-Hill Inc. Used with permission of McGraw-Hill Book Company.)

of oil pools to volume of sediments is found in the youngest group of oil-bearing sediments, the Pliocene rocks, deposited no later than 2.5 million years ago. It is also substantiated by the observation that the total amount of trapped oil decreases the further we move back in time (Fig. 7–10).

FIGURE 7–10 *Estimated production of crude oil from rocks of different geological age ranges.* (After G. C. Gester, 1948.)

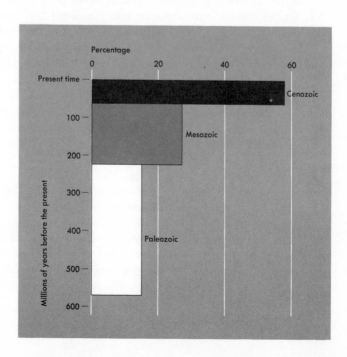

Natural gas, which is the lighter hydrocarbon fraction, and in particular methane, may range from a small quantity dissolved in the crude oil through a separate gaseous capping over an oil pool to a separate large accumulation not associated with a nearby oil pool. All such accumulations are valuable, and the technological mastery of pipeline-laying—and more recently of the commercial liquefaction of compressed natural gas—has made natural gas so widely available as a fuel that its use as an energy source in the United States now exceeds that of oil (Fig. 7–3).

Petroleum, like coal, has a geographic distribution that is widespread but uneven. The reasons for such a distribution are not obvious, as they are with coal; there is no known reason why rocks in Iran should be abundantly oil-bearing, whereas rocks of the same type and age in Australia should be barren. In fact, so uncertain is the question of geographic location that many authorities have reached the conclusion that the whole matter is only a reflection of the uneven intensity of oil search: Sufficient searching, they maintain, will reveal the same frequency of oil occurrence in all equivalent sedimentary basins, regardless of geography.

Production and consumption of petroleum have reached such tremendous proportions, for both the fuel and petrochemical industries, that it is difficult to believe that commercial production only began in 1857—little more than 100 years ago—in Rumania; this was followed two years later by production in the United States. The present leading production centers of the world (Fig. 7–11) are notably concentrated in a few countries.

What are the world's actual reserves of petroleum? Although the oil and gas industries maintain adequate measured and indicated reserves for near-term production, they lack a sure way of estimating the potential resources still in the ground. Oil and gas pools are small compared with coal fields and can be located only at high cost and with considerable difficulty. However, we know the dimensions of the sedimentary basins of the world, and if we assume that the oil potential in unexplored areas is just as good as it is in those most highly explored, such as the United States, it is possible to make a geological guess at the world's ultimate recoverable resources of oil and gas. Various authorities have done this and come up with answers that, so far from suggesting accord, differ by factors of three or four. Part of the uncertainty lies in the percentage of oil accepted as recoverable from an already located field. The present figure of 35 per cent recovery from the more efficient fields, for example, is considered too low by some, realistic by others. Another source of uncertainty lies in one's evaluation of just how well explored the sedimentary basins of a developed area like the United States actually are. Combining oil and gas (by taking 6,000 cubic feet of gas as equal to 1 barrel of crude oil), most estimates for the world's recoverable potential resources of petroleum

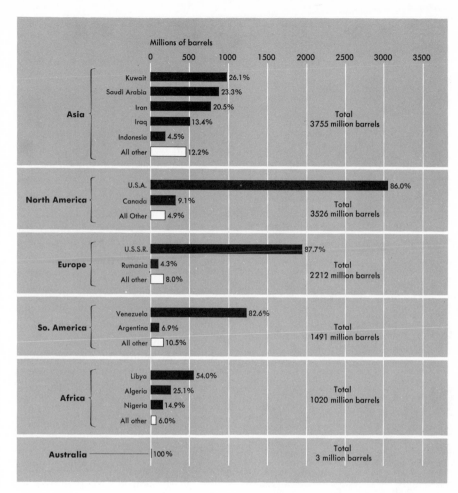

Millions of barrels

| | 0 | 500 | 1000 | 1500 | 2000 | 2500 | 3000 | 3500 |

Asia
- Kuwait — 26.1%
- Saudi Arabia — 23.3%
- Iran — 20.5%
- Iraq — 13.4%
- Indonesia — 4.5%
- All other — 12.2%

Total 3755 million barrels

North America
- U.S.A. — 86.0%
- Canada — 9.1%
- All Other — 4.9%

Total 3526 million barrels

Europe
- U.S.S.R. — 87.7%
- Rumania — 4.3%
- All other — 8.0%

Total 2212 million barrels

So. America
- Venezuela — 82.6%
- Argentina — 6.9%
- All other — 10.5%

Total 1491 million barrels

Africa
- Libya — 54.0%
- Algeria — 25.1%
- Nigeria — 14.9%
- All other — 6.0%

Total 1020 million barrels

Australia — 100%

Total 3 million barrels

FIGURE 7–11   *World distribution of the $12,890 \times 10^6$ barrels of petroleum produced in 1966. (After U.S. Bureau of Mines.)*

lie in the range 1,500–3,500 $\times$ 10^9 barrels.[4] The authoritative trade magazine "Oil and Gas Journal" in 1967 stated that combined oil and gas reserves for the world were 600 $\times$ 10^9 barrels. By this was meant only measured, indicated, and inferred reserves; but the estimates serve to emphasize that, as M. K. Hubbert has pointed out, our greatest period of oil discovery has quite possibly already been reached and that it is now time seriously to consider alternative sources of energy.

[4] The standard volume unit in crude oil production is the barrel, equal to 42 U.S. gallons. Because the specific gravity and calorific values of crude oils vary, it is possible to state only approximate conversion units to weight or energy values. 1 barrel = approximately 310 pounds and produces between 5,400,000 and 6,000,000 Btu of energy.

Tar Sands

One source of enormous size, known for years but only recently tapped, is tar sands. The greatest of all such deposits lies in northern Alberta, along the Athabasca River, and is estimated to contain the equivalent of 600×10^9 barrels of oil (Fig. 7–12). A tar sand is not unlike a crude oil, for it is a deposit of heavy large-molecule, asphaltic hydrocarbons. However, the tar does not run, flow, or migrate, but remains in place as a cementing agent for the mineral grains of the host sand. For the hydrocarbons to be recovered, the rock itself must be mined and heated with steam to make the asphalt flow; the resulting tarry extract must be processed to recover the valuable oil fraction. The Athabasca tar sands, whose origins are still uncertain, cover an area of 30,000 square miles and reach thicknesses of 200 feet.

Other tar sands and heavy viscous black oils similar to the tars are known but have not yet been seriously evaluated as resources. No individual deposits as large as the Athabasca sands have been found; nevertheless, the International Petroleum Encyclopedia for 1968 reports that estimates being made of the world's recoverable resources are as "large as those for conventional crude oil."

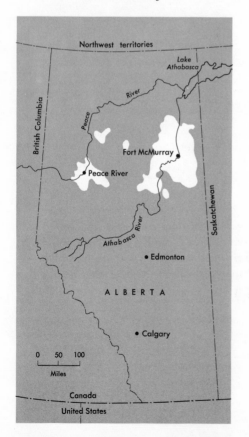

FIGURE 7–12 *Areas in Alberta, Canada, known to be underlain by tar sands. The first commercial production has begun near Fort McMurray, where 45,000 barrels of oil are produced from 100,000 tons of sand mined each day.*

Oil Shale

Some shales contain so much bituminous organic matter that significant quantities of petroleum can be extracted by direct distillation. Very locally, yields may reach four barrels of petroleum per ton of rock, but the richest deposits known—in Estonia—yield only one and one-half to two barrels per ton on a large scale.

The United States is fortunate in possessing the world's largest known reserves of oil shale. During the Eocene Epoch, three large shallow lakes existed in the intermontane area of Colorado, Wyoming, and Utah (Fig. 7–13); in them was deposited the series of rich organic shales and mudstone that are now the oil shales. Commonly known as the Green River Oil Shales, they are capable of producing between one-half and one and one-half barrels of oil per ton. Reserve estimates by the U.S. Geological Survey are enormous (850×10^9 barrels) and in excess of some of the more optimistic estimates of crude oil resources. Though minor scattered attempts at commercial exploitation of these huge resources have been attempted, they have not yet proven competitive with more easily won crude oil. Commercial exploitation of the Colorado shales, scheduled for the early 1970's, is the first serious attempt to recover petroleum products by distillation of a natural rock.

FIGURE 7–13 *Organic-rich shales were laid down in ancient freshwater lakes during the Eocene Epoch in parts of Colorado, Wyoming, and Utah. Little of these valuable sediments has been removed by later erosion. Although the whole area shown is underlain by oil shale, only the darkly shaded areas are underlain by shales that are more than 10 feet thick and that are capable of yielding 0.5 barrels or more oil per ton of shale. (After U.S. Department of Interior, 1968.)*

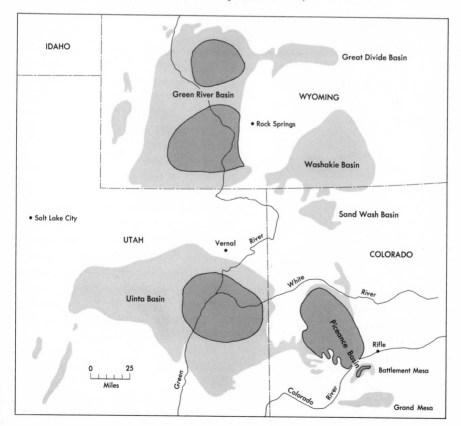

Oil shale resources in other parts of the world have not been adequately explored, but other huge deposits have already been found. The Irati Shales in Southeast Brazil, perhaps half as large as the oil shales in the United States, are the biggest units known outside North America. The 1967 world estimate for all known oil shales, including those in the United States, is placed at $2,000 \times 10^9$ barrels by the "Oil and Gas Journal." The figure may well grow as geologic exploration proceeds, but it is doubtful whether a recovery rate of better than 50 per cent can be achieved.

Comparison of Fossil Fuel Resources

The resource numbers mentioned have all been large (Table 7–2), but some experts would even suggest that they were conservative. Their size may lead one to a feeling of complacency, but this would be unfortunate: The world's demands for crude oil and natural gas have been almost doubling every ten years. It does not take much arithmetic to see that the present growth rate would seriously deplete even the most optimistically estimated petroleum resources by the end of the present century. Though cause for concern, the

Table 7–2

Potential Resources of Fossil Fuels and Their Energy*

	Barrels $\times 10^9$	Btu $\times 10^{15}$
Coal	34,500	197,000
Oil and Gas	2,500	14,250
Oil Shale	1,000	5,700
Tar Sands	at least 600	3,400
1966 Energy Consump.	equiv. to 30	170
1976 Energy Consump.	(estimated) 48	270

* 1 barrel of crude oil is taken to be approximately equal to 0.22 tons of coal and to generate 5.7×10^6 Btu of energy.

situation is probably not cause for alarm. In the first place, the growth rate is unlikely to continue; in the second, assuming that the efficient oil industry will find, develop, and extract most of the presently undiscovered or un-recovered oil that geologists now say exists, man has a transition period ahead when other fossil fuels and the nonfossil fuel energy sources can be developed. The beginnings of this trend are already apparent in the interest that some governments and major oil companies are showing in oil shale, tar sands, and coal.

Nuclear Fission Energy

Nuclear energy is derived from either of two different processes. The fissioning of certain atomically heavy radioactive elements produces two or more lighter elements and releases energy in the process. Similarly, fusion of two or more very light elements to produce heavier elements also releases energy. The fission process is that used in the atom bomb and can be controlled to release power in an orderly manner. Fusion is the hydrogen bomb reaction and has not yet been controlled for orderly power production. Because fusion power still requires a massive technological advance, we will confine our discussion to the source of fission power, which is already capable of production.

The only naturally occurring fissionable atom is the uranium-235 isotope, which comprises 0.7 per cent of all natural uranium. When U^{235} absorbs a neutron, it becomes unstable and fissions into lighter elements, giving off more neutrons and heat in the process. If the daughter neutrons are made to generate further fissions, a sustained chain reaction follows and the heat produced can be used to generate electrical energy. The U^{235} chain reaction was first achieved by Professor Enrico Fermi on December 2, 1942, in what has become one of the most important experiments ever performed.

Uranium-235 is scarce and the cost of concentrating it from raw uranium high, but the more abundant isotope, U^{238}, can be converted to a fissionable atom of plutonium-239 by addition of neutrons in a process known as *breeding* (Fig. 7–14). Similarly, the common isotope of thorium, Th^{232}, can be converted to fissionable U^{233} in a breeder reaction.

Most of the work on atomic energy has centered on the uranium reactions, and it is the uranium resources that have been most actively exploited and

FIGURE 7–14 *Schematic diagram of the controlled U^{235} reaction and the U^{238} breeder reaction to produce fissionable Pu^{239}. The heat generated can be converted to electricity.*

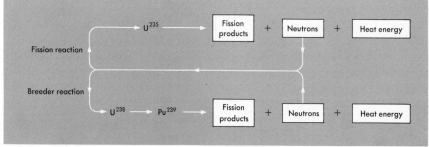

explored. Fissioning of one pound of uranium produces heat energy equivalent to approximately 5,900 barrels of crude oil, so even moderate reserves of uranium lead to tremendous energy resources.

Uranium is a geochemically scarce metal, but because it is used mainly as a fuel it is discussed here rather than in Chapter 4 with the other scarce metals. It occurs in two valence states, U^{+4} and U^{+6}, and the interplay of these governs its distribution in the crust. In silicic igneous rocks, pegmatites, and magmatic hydrothermal veins, uranium is widespread as the mineral uraninite (sometimes called pitchblende) which has uranium in the U^{+4} state. The U^{+4} state is readily oxidized to U^{+6} under conditions at the Earth's surface, in which form it will combine with oxygen to form the complex uranyl ion $(UO_2)^{+2}$ which itself has a positive valence and can form separate compounds. The uranyl ion forms soluble compounds, such as uranyl carbonate, which facilitate the movement of uranium in surface waters. Precipitation occurs when the solutions encounter a reducing agent, such as organic matter, that returns the uranium to the less soluble U^{+4} state. The reduced uranium may precipitate as uraninite, as it has done in many of the famous Colorado Plateau deposits where buried logs have been found almost completely replaced by uraninite; or it may precipitate as uranium-rich organic compounds, as it has done in many of the organic-rich sediments around the world, such as the 40,000 square miles of Chattanooga Shale in Alabama and Kentucky, and in the Alum Shale of Sweden and Norway. Finally, the uranium may precipitate as a substituting element in a carrier mineral, and the common mineral in which this occurs is apatite, $Ca_5(PO_4)_3(OH,F)$, with U^{+4} atoms replacing some of the Ca atoms, a form in which it apparently occurs in the Florida phosphate deposits.

The measured resources of uranium in rich deposits are unfortunately not so large as we might hope. During the years 1945–1960 uranium received the most intensive mineralogical and geological scrutiny ever accorded a metal. A great many rich deposits were located and—following the assumption that further prospecting would locate yet more high-grade deposits—confident and rosy predictions were made of large, easily won potential resources. Further work suggests that this confidence was misplaced.

The richest deposits in the United States are apparently secondary concentrations in sedimentary rocks, arising from the movement of uranium by ground water. The largest and most numerous deposits are found in rocks of the Jurassic and Triassic Epochs on the Colorado Plateau—in western Colorado, eastern Utah, northeastern Arizona, and northwestern New Mexico. Other rich deposits in sedimentary rocks, but from the Cenozoic Epochs, are found in Wyoming. All of these deposits, together with smaller and less valuable occurrences in numerous other states, account for a measured reserve of uranium reported by the United States Atomic Energy Commission to be only 145,000 tons (Table 7–3). The larger measured reserves reported for Canada are principally located in rich hydrothermal vein deposits in the Great Bear Lake region, Northwest Territories, and in the Blind River district, north of Lake Huron in Ontario, where extensive conglomerate beds of Precambrian age contain disseminated uranium minerals of uncertain origin. The large South African reserves are principally associated with the Witwatersrand gold deposits, where trace amounts of uranium are recovered as a by-product from gold production.

The measured reserves of uranium in the non-Communist world are not large. Because uranium is such a strategic commodity, however, figures such as those in Table 7–3 should be accepted with doubt, for an element of caution, even secrecy, surrounds them. The figures are certainly conservative. They reveal only part of the story, however. With a price of $8 per pound for U_3O_8 —the form in which uranium concentrates are usually sold—only the richest ores can be profitably worked, and hence these are the only ones included in the reserves. Because U^{235} is the isotope currently used for power production, the high cost of the uranium fuel becomes a limiting factor in the cost of power produced. If breeder reactors, using abundant U^{238}, can be perfected, the cost of fuel as a consequence will be much less critical, with the result that more abundant but lower-grade uranium resources can then be considered.

Table 7–3

Measured Reserves of Uranium in the Non-Communist World

Country	Uranium (Tons)
Canada	180,000
United States	145,000
Republic of So. Africa	130,000
France	32,000
India	9,500
Australia	8,500
All others	25,000

U.S. Bureau of Mines, 1965.

The United States is particularly well endowed with lower-grade potential resources, principally in widespread organic-rich or phosphatic sediments, but also in presently uneconomic deposits in the Colorado Plateau (Fig. 7–15). The question of when such resources might become reserves depends largely on the speed with which the technological problems of breeder plants are overcome. Anticipating this day, the United States Atomic Energy Commission estimated, in 1963, how increasing the price of U_3O_8 might increase the reserves of available uranium (Table 7–4). Even these figures are probably now conservative, but they serve to illustrate the tremendous power potential that lies in uranium for the centuries ahead. For example, the 20,000,000 tons of uranium predicted as recoverable in a price range of $30–$100 per pound for U_3O_8 could produce power equivalent to that from $250,000 \times 10^9$ barrels

Energy: the saga of the fossil fuels

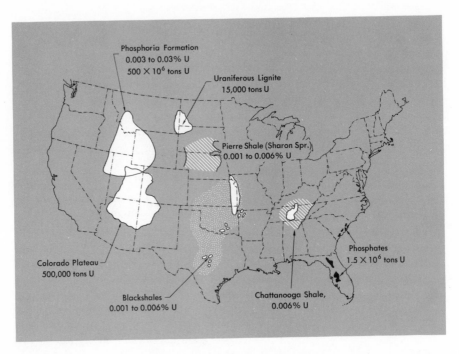

FIGURE 7–15 *The major potential resources of uranium in the United States occur in low-grade deposits. Tonnage estimates have been made for the Colorado Plateau, Florida Phosphate, and Phosphoria Formation deposits, but even larger quantities are available in some of the uraniferous organic-rich shales. (After M. K. Hubbert,* Energy Resources, *Pub. 1000-D, Committee on Natural Resources, Nat. Academy of Sciences—Nat. Research Council, Washington, D.C., 1962.)*

Table 7–4

Potential Resources of Uranium
in the U.S. and Price per Pound
as Reserves

Price per Pound of U_3O_8	Potential Resources (Tons)
Less than $10	650,000
$10 to $30	600,000
$30 to $100	20,000,000
$100 to $500	1,700,000,000

After U.S. Atomic Energy Commission, 1963.

of crude oil—approximately seven times the estimated energy from the world's fossil fuel supply.

Realizing its tremendous potential, both private and public sectors have tried to develop nuclear-powered electrical-generating stations, but on a competitive basis with other energy sources. The goal has not yet been reached, but the increasing number of nuclear power plants in the United States alone suggests that it will be done within a very few years and, further, that nuclear energy may well produce as much as 5 per cent of the country's power needs by 1980.

Energy: the saga of the fossil fuels

Other Energy Resources

Evidence has been rapidly accumulating in recent years to suggest that the tremendous burning rate of fossil fuels is seriously contaminating the Earth's atmosphere with its principal product of combustion, carbon dioxide. The long-term effects of such a contamination could lead to major ecological and climatological changes on Earth, for carbon dioxide is one of the most important gases in the atmosphere's thermal regulation system. Similarly, concern has been repeatedly expressed that disposal of the highly radioactive and deadly waste products of fission reactions constitutes a difficulty in the widespread development of nuclear power. Such concerns have given added strength to drives to develop other sources of power.

The only important alternative source appears to be electrical generation from water power. Water power involves a problem of production rate rather than of total energy, for the sun's energy drives the hydrological cycle continuously, and water reservoirs are replenished at the same rate. Questions arise: How much power can be supplied on a continuous basis? Can it be done close enough to power-consumption sites to be worth while? Adams estimated in 1961 that the ultimate water power capacity of the United States is already 23 per cent developed, suggesting that the total annual water power capacity of the United States has a crude oil combustion equivalent of about only 30 × 10⁹ barrels. Like other energy sources, water power is unevenly distributed around the world (Table 7–5), and although it is vitally important where available, it does not have the unused capacity needed to replace the depleting fossil fuel resources.

Table 7–5

Distribution of Potential Water-power Resources

Country	Percentage of Total
North America	11
South America	20
Western Europe	6
Africa	27
South East Asia	16
Middle East, Far East, Australia	4
U.S.S.R. and China	16

After Adams, quoted by M. K. Hubbert, 1962.

The largest untapped power source—as is clear from Fig. 7–1—is the solar radiation energy falling on the Earth. Many authorities view this as the long-term power source to which man must ultimately turn. How the solar energy is to be captured and converted to more useful energy forms, such as electricity, remains so problematical that separate discussion of the many imaginative schemes proposed is not warranted. There can be little doubt, however, that sources of energy that can be considered potential resources do exist, and that they are so large that our future energy problems will probably be technological in nature, rather than problems of abundance.

8

Water

. . . pure water is becoming a critical commodity whose abundance is about to set an upper limit of economic evolution in a few parts of the Nation and inevitably will do so rather widely within half a century or less. Prudence requires that the Nation learn to manage its water supplies boldly, imaginatively, and with utmost efficiency. Time in which to develop such competence is all too short. (A. M. Piper, U.S. Geological Survey, Water-Supply Paper #1797, 1965.)

Water is such a vital and essential resource that to attempt to assign it a cash value would be pointless; simply put, it is the most valuable of all our resources.

Although much water is locked in the minerals of the crust, such as that in muscovite, $KAl_3(Si_3O_{10})(OH)_2$, it is the free water of the hydrosphere from which man must draw his resources. The total amount of water in the hydrosphere is estimated to be 326×10^6 cubic miles, or 359×10^{18} gallons, but it is very unevenly distributed (Fig. 8–1), with 97.2 per cent residing in the oceans and 2.15 per cent of the remainder trapped in the polar icecaps and glaciers. Except as transportation media, these two largest water reservoirs are of little use as major resources; the seas are too saline, the icecaps too

	Location	Water volume, gallons	Percentage of total water
Surface water			
	Fresh-water lakes	33×10^{15}	.009
	Saline lakes and inland seas	27.5×10^{15}	.008
	Average in stream channels	0.3×10^{15}	.0001
Subsurface water			
	Vadose water (includes soil moisture)	17.6×10^{15}	.005
	Ground water within depth of half a mile	1100×10^{15}	.31
	Ground water—deep lying	1100×10^{15}	.31
Other water locations			
	Icecaps and glaciers	7700×10^{15}	2.15
	Atmosphere	34.1×10^{15}	.001
	World ocean	$348,700 \times 10^{15}$	97.2

FIGURE 8-1   *Distribution of water in the hydrosphere. (After U.S. Geological Survey.)*

remote, and the costs of conversion and recovery too high to warrant any but local use. The remaining 0.65 per cent of the water in the hydrosphere is the fraction we must rely on for man's use, and this small portion would soon be consumed were it not for the well-known hydrological cycle (Fig. 8-2). The solar-energy-driven cycle of evaporation plus transpiration followed by condensation, then precipitation, assures a continuous supply and makes water a renewable resource. The problems of water are therefore not only to do with abundance, but also with distribution and rates of supply. Some areas are well supplied, others water-poor. More than any other factor, availability of water determines the ultimate population capacity of a geographic province.

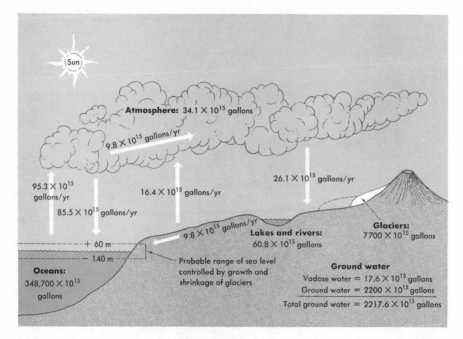

FIGURE 8–2 *The hydrosphere and the hydrological cycle. (After A. L. Bloom,* The Surface of the Earth, *Prentice-Hall, 1969.)*

Distribution of Precipitated Water

The country for which most information is available on the details of the hydrological cycle is the United States. Each year an average of 4.75×10^9 acre feet,[1] or $1,552 \times 10^{12}$ gallons, of water fall as rain. The average rainfall is not evenly distributed, however, and precipitation may vary by amounts of 20 times or more across large geographic provinces (Fig. 8–3). The regional disparities that develop as a result of this are perhaps best demonstrated by the fact that the area east of the Mississippi receives 65 per cent of the country's total rainfall (excluding Alaska and Hawaii), whereas that to the west receives only 35 per cent.

In addition to geographic variations, there are marked temporal variations in precipitation. Seasonal variations are, of course, obvious to all, but in addition there are considerable long-term fluctuations in precipitation related to long-term weather cycles (Fig. 8–4). Although water is a renewable resource, a realistic evaluation of the supply rate must necessarily be made over a time interval of many years.

---

[1] An acre foot is a commonly used measure of rainfall and is the amount of water that will cover an acre of land to a depth of one foot. It is equivalent to 326,700 gallons.

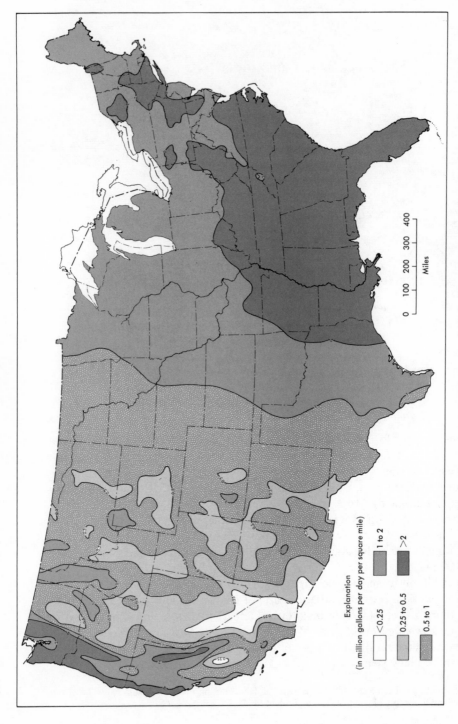

Explanation
(in million gallons per day per square mile)

☐ <0.25	☐ 1 to 2
☐ 0.25 to 0.5	■ >2
☐ 0.5 to 1	

0   100   200   300   400

Miles

FIGURE 8–3 *The average rainfall in the United States—excluding Hawaii and Alaska—illustrating the tremendous regional variations that occur. (After A. M. Piper, U.S. Geological Survey, Water-Supply Paper #1797, 1965.)*

Evaporation and Transpiration

Water will *evaporate* from any open-water body and from any wetted surface. A significantly large fraction of the rainfall is returned to the atmosphere in this fashion. In addition, water is assimilated by the root systems of growing plants and is later *transpired* from the leaf surfaces by a process essentially identical to evaporation. The two effects, evaporation and transpiration, cannot be individually discriminated for their effectiveness in returning

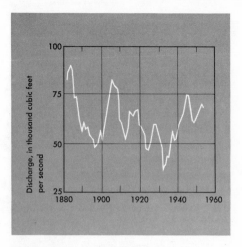

FIGURE 8–4 *Variations in the water discharge of the Mississippi River, at Keokuk, Iowa, reflecting regular long-term variations in the rainfall and climate of the Mississippi River's drainage basin. (After U.S. Geological Survey.)*

rainfall to the atmosphere, but their sum contribution can be evaluated and is usually called the *evapotranspiration* factor. The fraction of rain falling on the United States that is returned to the atmosphere by evapotranspiration, for example, is 70 per cent; for the world as a whole, approximately 62 per cent (see Fig. 8–2). In more arid countries, such as Australia, the fraction is larger, and in those less arid, such as the United Kingdom, it is lower. Water returned to the atmosphere by evapotranspiration is unavailable to man, except in the sense that useful plants may be grown in the place of useless ones. It cannot be trapped and redistributed for industrial or other purposes.

In regions of low rainfall, plant cover will develop to a point where all precipitation is used in evapotranspiration and none remains for stream flow. Clearly, the seasonal rainfalls provide a qualifier for this, and in periods of rainfall maxima streams will flow even in the most arid areas. In general, if the potential evapotranspiration—that which would result from the maximum plant-cover a region could carry under ideal circumstances—should exceed the precipitation, overland stream flow ceases. Conversely, if evapotranspiration is less than precipitation, runoff is generated perennially.

The amount by which precipitation exceeds evapotranspiration is clearly the perennial yield of stream flow water, and this is withdrawable and hence the usable fraction. Across the entire United States the withdrawable water amounts to 30 per cent of the total rainfall, or approximately 466×10^{12} gallons per year. If we consider the distribution of water deficiencies and surpluses (Fig. 8–5), it is clear that essentially the entire eastern half of the United States,

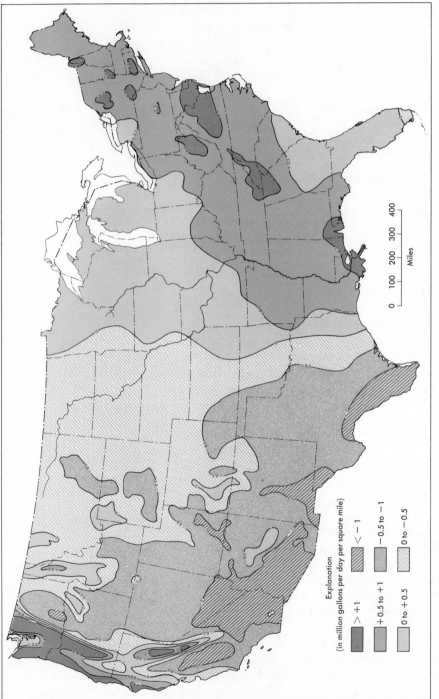

FIGURE 8-5   The balance between potential evapotranspiration and rainfall precipitation. In regions where rainfall exceeds evapotranspiration, there is a water surplus that appears as a perennial stream flow. When potential evapotranspiration exceeds rainfall, there is a net water deficiency. (After A. M. Piper, U.S. Geological Survey, Water-Supply Paper #1797, 1965.)

Explanation
(in million gallons per day per square mile)

> +1

+0.5 to +1

0 to +0.5

< −1

−0.5 to −1

0 to −0.5

0   100   200   300   400
            Miles

together with a small region in the Pacific northwest, enjoys water surplus, while most of the West is water-deficient and arid.

Ground Water

Though rainfall is the only water supply on a long-term basis, there is a very important temporary reservoir available to man. Water in any form that occurs in the ground is commonly called *subsurface water*. It is one of the most important water resources, but unfortunately its rate of replenishment is often so slow that overly rapid withdrawal can cause serious local depletion. Beneath the land areas of the world there is a zone where all rock pores and openings are saturated with water. This is *ground water*, and the upper surface of the saturated zone is the water table. The water table may lie at the surface, as in a lake or stream, or hundreds of feet below the surface; however, it is always present. Water below the water table is moving by seepage and slow flow to the sea (see Fig. 8-2); the ground water region can thus be considered analogous to a large but very slow river in which residence times for ground water to reach the sea may vary from hours to thousands of years. The water table is not horizontal, but because of the varying resistance offered by rocks to the flow of ground water, it is irregular and tends to reflect the topography above (Fig. 8-6).

FIGURE 8–6   *Relation between subsurface waters, water table, and topography.*

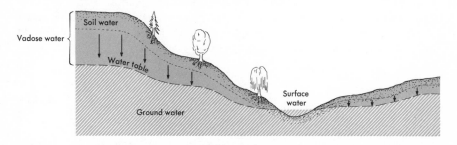

Above the water table is an unsaturated region of *vadose water*. Near the surface, where plant roots are abundant, is the region of *soil water*, in which water is moving neither up nor down, but adhering to the surface of mineral grains. Somewhat deeper, but still in the unsaturated zone, vadose water is slowly seeping down to join the body of ground water below. Vadose waters cannot be considered direct resources as can ground waters, but it is the vadose waters that serve to replenish—albeit slowly—any withdrawal from the ground water zone.

The amount of ground water is huge (Fig. 8–1), an estimated 3,000 times larger than the volume of water in all the rivers at any given time. The major problems in its exploitation are three-fold. First, where rock porosity is very low and permeability poor, so slow is the flow of water into wells that water cannot be removed at a worthwhile rate. Adequate recovery thus requires a suitable aquifer, or water carrier, as a supplier. Second, the rate of replenishment is slow because much of the supply—rainfall—runs off in rivers. It has been estimated, for example, that the total ground water resources in the United States to a depth of 2,500 feet would take an average of 150 years to be recharged if they were all removed—although some recharging areas would of course be slower than others. Thus, if we pump water from the ground at a rate greater than that of replenishment, we are essentially mining the water. This problem has now become acute in parts of the arid Southwestern United States, where withdrawal at rates of up to a hundred times greater than those of replenishment has unfortunately been practised in some of the richer ground water zones; replenishment, assuming cessation of all pumping, would take periods of up to 100 years or more. Finally, there is the problem of water quality. As ground water moves through the rocks, it dissolves the more soluble constituents. The problem varies with host rock, water depth, flow rates, and other factors, but in general a water with more than 0.05 per cent (500 parts per million) of dissolved salts is unsuitable for human consumption, and one with more than 0.2 per cent dissolved salts is unsuitable for almost all other uses; however, waters as saline as 1.0 per cent can be used for some special purposes. The portion of the United States underlain by aquifers that will yield good-quality water at flow rates of 50 gallons per minute or more is extensive (Fig. 8–7), and wise development and utilization of these resources is now the prime concern of many able scientists and engineers.

Rates of Use and Reserves

In 1965, William H. Kerns, a water specialist working for the U.S. Department of the Interior, cogently stated that ". . . all parts of the [United States] either have or will have water problems. The well-watered Eastern and Southern States are beginning to share a concern about water that has been felt in the arid West since its settlement. As industrial development and urbanization expand in the East, it is becoming more apparent each year that lack of water may deter growth unless action is taken to assure a continued supply. "

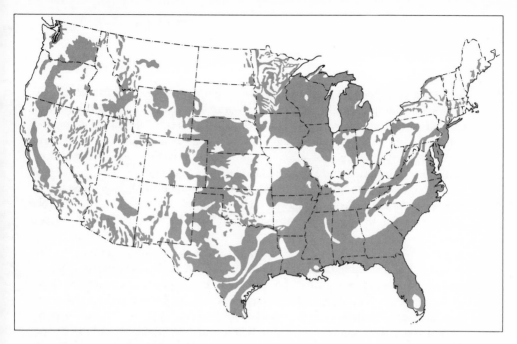

FIGURE 8–7 *Regions of the United States underlain by aquifers capable of yielding water containing 0.2 per cent dissolved solids or less at rates of 50 gallons per minute from individual wells. (After H. E. Thomas, 1955.)*

The current usage of rainfall across the United States is summarized in Fig. 8–8. Approximately 8 per cent of the rain falling on the United States, or 27 per cent of the stream flow—amounting to 110×10^{12} gallons—is withdrawn for use in the categories shown. Of course, not all of the water withdrawn

FIGURE 8–8 *Generalized distribution and use of the annual precipitation in the United States. (After A. Wolman,* Water Resources, *Pub. 1000-B, Committee on Natural Resources, Nat. Academy of Sciences—Nat. Research Council, Washington, D.C., 1962.)*

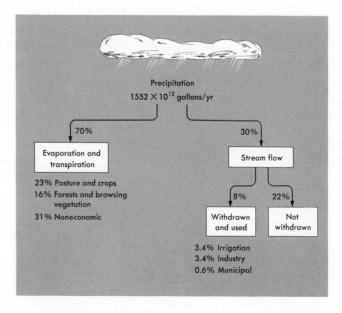

for use is consumed in a single use cycle. Water used to cool industrial machines, for example, can be returned to streams and reused. Irrigation and other agriculture water, however, is very largely consumed by evapotranspiration and hence is lost; it has been estimated by MacKichan and Kammerer that approximately 25 per cent of all water now withdrawn for use in the United States is so consumed and thus unavailable for further use.

Taking account of the fraction of water consumed, and projecting future population growth, Wollman has estimated that by 1980, use of withdrawable water must rise to 51 per cent of the stream flow; by the year 2000, to 81 per cent.

Stream flow is temporal, but man's water demands are constant. We are forced to build suitable damming, storage, and reticulation schemes so that excess flows of good seasons can be conserved to supplement diminished flows of the poor. In so doing, of course, large open water bodies such as reservoirs and lakes are created, and evaporation loss inevitably rises. Though experts differ, it has been optimistically stated that as much as 90 per cent of the stream flow water may eventually be utilized. Even this optimistic estimate—which is made under the assumption that the necessary engineering and political problems could all be overcome—is dangerously close to Wollman's predicted withdrawal needs by the year 2000. The sense of urgency expressed by Dr. Piper is very real.

All of the projections and discussions made for water use are made under the assumption that adequate stream water purity is maintained. The seriousness of the problem of water pollution, a subject widely discussed today, is self-evident. As the percentage of stream flow withdrawn for use rises, it becomes less and less feasible to use the same stream channels for effluent disposal, for a small quantity of polluted water can cause a whole stream to become polluted and hence unusable. The world's growing population is bringing with it a water problem fast enough, and to compound the problem by failure to control pollution, and hence to squander our most valuable natural resource, would be foolhardy in the extreme.

Appendix

Table 1

The Chemical Elements and Their Crustal Abundances

Name	Symbol	Atomic Number	Crustal Abundance, Weight Percent
Actinium	Ac	89	Man-made
Aluminum	Al	13	8.00
Americium	Am	95	Man-made
Antimony	Sb	51	0.00002
Argon	Ar	18	Not known
Arsenic	As	33	0.00020
Astatine	At	85	Man-made
Barium	Ba	56	0.0380
Berkelium	Bk	97	Man-made
Beryllium	Be	4	0.00020
Bismuth	Bi	83	0.0000004
Boron	B	5	0.0007
Bromine	Br	35	0.00040
Cadmium	Cd	48	0.000018
Calcium	Ca	20	5.06
Californium	Cf	98	Man-made
Carbon[a]	C	6	0.02[d]
Cerium	Ce	58	0.0083
Cesium	Cs	55	0.00016
Chlorine	Cl	17	0.0190
Chromium	Cr	24	0.0096
Cobalt	Co	27	0.0028
Copper	Cu	29	0.0058

Table 1 (cont.)

Name	Symbol	Atomic Number	Crustal Abundance, Weight Percent
Curium	Cm	96	Man-made
Dysprosium	Dy	66	0.00085
Einsteinium	Es	99	Man-made
Erbium	Er	68	0.00036
Europium	Eu	63	0.00022
Fermium	Fm	100	Man-made
Fluorine	F	9	0.0460
Francium	Fr	87	Man-made
Gadolinium	Gd	64	0.00063
Gallium	Ga	31	0.0017
Germanium	Ge	32	0.00013
Gold	Au	79	0.0000002
Hafnium	Hf	72	0.0004
Helium	He	2	Not known
Holmium	Ho	67	0.00016
Hydrogen[b]	H	1	0.14
Indium	In	49	0.00002
Iodine	I	53	0.00005
Iridium	Ir	77	0.00000002
Iron	Fe	26	5.80
Krypton	Kr	36	Not known
Lanthanum	La	57	0.0050
Lead	Pb	82	0.0010
Lithium	Li	3	0.0020
Lutetium	Lu	71	0.000080
Magnesium	Mg	12	2.77
Manganese	Mn	25	0.100
Mendelevium	Md	101	Man-made
Mercury	Hg	80	0.000002
Molybdenum	Mo	42	0.00012
Neodymium	Nd	60	0.0044
Neon	Ne	10	Not known
Neptunium	Np	93	Man-made
Nickel	Ni	28	0.0072
Niobium	Nb	41	0.0020
Nitrogen	N	7	0.0020
Nobelium	No	102	Man-made

Table 1 (cont.)

Name	Symbol	Atomic Number	Crustal Abundance, Weight Percent
Osmium	Os	76	0.00000002
Oxygen[b]	O	8	45.2
Palladium	Pd	46	0.0000003
Phosphorus	P	15	0.1010
Platinum	Pt	78	0.0000005
Plutonium	Pu	94	Man-made
Polonium	Po	84	Footnote[d]
Potassium	K	19	1.68
Praseodymium	Pr	59	0.0013
Promethium	Pm	61	Man-made
Protactinium	Pa	91	Footnote[d]
Radium	Ra	88	Footnote[d]
Radon	Rn	86	Footnote[d]
Rhenium	Re	75	0.00000004
Rhodium[c]	Rh	45	0.00000001
Rubidium	Rb	37	0.0070
Ruthenium[c]	Ru	44	0.00000001
Samarium	Sm	62	0.00077
Scandium	Sc	21	0.0022
Selenium	Se	34	0.000005
Silicon	Si	14	27.20
Silver	Ag	47	0.000008
Sodium	Na	11	2.32
Strontium	Sr	38	0.0450
Sulfur	S	16	0.030
Tantalum	Ta	73	0.00024
Technetium	Tc	43	Man-made
Tellurium[c]	Te	52	0.000001
Terbium	Tb	65	0.00010
Thallium	Tl	81	0.000047
Thorium	Th	90	0.00058
Thulium	Tm	69	0.000052
Tin	Sn	50	0.00015
Titanium	Ti	22	0.86
Tungsten	W	74	0.00010
Uranium	U	92	0.00016
Vanadium	V	23	0.0170

Table 1 (cont.)

Name	Symbol	Atomic Number	Crustal Abundance, Weight Percent
Xenon	Xe	54	Not known
Ytterbium	Yb	70	0.00034
Yttrium	Y	39	0.0035
Zinc	Zn	30	0.0082
Zirconium	Zr	40	0.0140

After K. K. Turekian, (1969).

[a] Estimate from S. R. Taylor (1964).

[b] Analyses of crustal rocks do not usually include separate determinations for hydrogen and oxygen. Both combine in essentially constant proportions with other elements, so abundances can be calculated.

[c] Estimates are uncertain and have a very low reliability.

[d] Elements formed by radioactive decay of long-lived uranium and thorium. The daughter products are themselves radioactive but have such short half-lives that their crustal accumulations are too low to be measured accurately.

Table 2

Principal Ore Minerals and Annual World Production Rate (1967) of the Technologically Important Metals

I. THE GEOCHEMICALLY ABUNDANT METALS

Elements	World Production (Tons)	Principal Ore Minerals
Iron	618,308,000 (iron ore)	Magnetite, Fe_3O_4; hematite, Fe_2O_3; goethite, $HFeO_2$; siderite, $FeCO_3$
Aluminum	8,285,000 (metal)	Gibbsite, H_3AlO_3; diaspore, $HAlO_2$; boehmite, $HAlO_2$; kaolinite, $Al_2Si_2O_5(OH)_4$; anorthite, $CaAl_2Si_2O_8$
Chromium	5,111,000 (chromite)	Chromite, Fe_2CrO_4
Titanium	3,278,000 (titanium ore)	Rutile, TiO_2; ilmenite, $FeTiO_3$
Manganese	18,650,000 (manganese ore)	Pyrolusite, MnO_2; psilomelane, $BaMn_9O_{18} \cdot 2H_2O$; cryptomelane, KMn_8O_{16}; rhodocrosite, $MnCO_3$
Magnesium	202,600 (metal), 9,947,000 (carbonate)	Magnesite, $MgCO_3$; dolomite, $CaMg(CO_3)_2$

Table 2 (cont.)

II. THE GEOCHEMICALLY SCARCE METALS

A. Metals Commonly Forming Sulfide Minerals

Elements	World Production (Tons)	Principal Ore Minerals
Copper	5,406,000	Covellite, CuS; chalcocite, Cu_2S; digenite, Cu_9S_5; chalcopyrite, $CuFeS_2$; bornite, Cu_5FeS_4; tetrahedrite, $Cu_{12}Sb_4S_{13}$
Zinc	5,175,000	Sphalerite, ZnS
Lead	3,133,000	Galena, PbS
Nickel	481,000	Pentlandite, $(Ni,Fe)_9S_8$; Garnierite, $H_4Ni_3Si_2O_9$
Antimony	64,400	Stibnite, Sb_2S_3
Molybdenum	62,680[a]	Molybdenite, MoS_2
Arsenic	greater than 40,000	Arsenopyrite, $FeAsS$; orpiment, As_2S_3; realgar, AsS
Cadmium	14,300[b]	Substitution for Zn in sphalerite
Cobalt	20,000	Linnaeite, Co_3S_4; substitution for Fe in pyrite, FeS_2
Mercury	9,200	Cinnabar, HgS; metacinnabar, HgS
Silver	8,970	Acanthite, Ag_2S; substitution for Cu and Pb in their common ore minerals
Bismuth	3,460	Bismuthinite, Bi_2S_3

B. Metals Commonly Found in the Native Form

Elements	World Production (Tons)	Other Important Ore Minerals Besides the Native Elements
Gold	1569	Calaverite, $AuTe_2$; krennerite, $(Au,Ag)Te_2$, sylvanite, $AuAgTe_4$; petzite, $AuAg_3Te_2$
Platinum	44[c]	Sperrylite, $PtAs_2$; braggite, PtS_2; cooperite, PtS
Palladium	43[c]	Arsenopalladinite, Pd_3As; michenerite, $PdBi_2$; froodite, $PdBi_2$
Rhodium	9.6[c]	———
Iridium	6.4[c]	———
Ruthenium	4.4[c]	Laurite, RuS_2
Osmium	1.1[c]	———

Table 2 (cont.)

C. Metals Commonly Forming Oxygen-containing Compounds

Elements	World Production (Tons)	Principal Ore Minerals
Tin	211,664	Cassiterite, SnO_2
Tungsten	30,430	Wolframite, $FeWO_4$; Scheelite, $CaWO_4$
Uranium[a]	14,500	Uraninite (pitchblende), UO_2
Vanadium	10,600	Carnotite, $K_2(UO_2)_2(VO_4)_2 \cdot 3H_2O$ Substituting for Fe in magnetite, Fe_3O_4
Niobium[ab]	5,500	Columbite, $FeNb_2O_6$; Pyrochlore, $NaCaNb_2O_6F$
Tantalum[ab]	720	Tantalite, $FeTa_2O_6$
Beryllium	180	Beryl, $Be_3Al_2(SiO_3)_6$

After U.S. Bureau of Mines.

[a] Figures known for the Western-bloc nations only.

[b] Estimate for 1966.

[c] Estimated by dividing the platinoid metal production of 108.5 tons in proportion to the relative metal abundances.

Suggestions for further reading

General

Flawn, P. T., *Mineral Resources*. Chicago: Rand McNally, 1966.

Krauskopf, K. B., *Introduction to Geochemistry*. New York: McGraw-Hill, 1967.

Landsberg, H. H., *Natural Resources for U.S. Growth: A Look Ahead to the Year 2000*. Baltimore: Johns Hopkins, 1964.

Lovering, T. S., *Minerals in World Affairs*. Englewood Cliffs, N. J.: Prentice-Hall, 1943.

Mason, B., *Principles of Geochemistry*. New York: Wiley, 1966.

McDivitt, J. F., *Minerals and Men*. Baltimore: Johns Hopkins, 1965.

Mero, J. L., *The Mineral Resources of the Sea*. Amsterdam: Elsevier, 1965.

The Staff, *Mineral Facts and Problems*. Bull. 630, U.S. Bureau of Mines, 1965.

The Staff, *Minerals Yearbook, Volumes I and II: Metals, Minerals and Fuels*. U.S. Bureau of Mines, published annually.

Metal deposits

Bateman, A. M., *The Formation of Mineral Deposits*. New York: Wiley, 1951.

Park, C. F., Jr., and R. A. MacDiarmid, *Ore Deposits*. San Francisco: Freeman, 1964.

Routhier, P., *Les Gisements Métallifères, Tomes I et II*. Paris: Masson, 1963.

Titley, S. R., and C. L. Hicks, *Geology of the Porphyry Copper Deposits of Southwestern North America*. Tucson: Univ. Ariz. Press, 1966.

Industrial minerals

Bates, R. L., *Geology of the Industrial Rocks and Minerals*. New York: Harper, 1960.

Borchert, H., and R. O. Muir, *Salt Deposits— The Origin, Metamorphism, and Deformation of Evaporites*. Princeton: Van Nostrand, 1964.

Gillson, J. L., and others, *Industrial Minerals and Rocks*. Seeley W. Mudd Series, 3rd ed., American Institute of Mining, Metallurgical and Petroleum Engineers, 1960.

Lotze, F., *Steinsalz und Kalisalz*, 2nd ed. Berlin-Nikolassee: Borntraeger, 1957.

Water resources

Davis, S. N., and R. J. M. DeWiest, *Hydrogeology*. New York: Wiley, 1966.

DeWiest, R. J. M., *Geohydrology*. New York: Wiley, 1965.

Fox, C. S., *The Geology of Water Supply*. London: London's Technical Press, 1949.

McGuinness, C. L., *The Role of Ground Water in the National Water Situation*. U.S. Geol. Survey, Water-Supply Paper #1800, 1963.

U.S. Department of Agriculture, *Water. The Yearbook of Agriculture, 1955*.

White, D. E., J. D. Hem, and G. A. Waring,

Chemical Composition of Subsurface Waters. U.S. Geol. Survey Prof. Paper 440-F, 1963.
Wolman, A., *Water Resources,* Publication 1000-B. Nat. Acad. Sciences—Nat. Research Council, 1962.

Fuel resources

Averitt, P., *Coal Resources of the United States Jan. 1, 1967.* U.S. Geol. Surv. Bull. 1275.
International Petroleum Encyclopedia. Tulsa: Petroleum Pub. Co., 1968.
Levorsen, A. I., *Geology of Petroleum,* 2nd ed. San Francisco: Freeman, 1967.
Nagy, B., and U. Colombo (*eds.*), *Fundamental Aspects of Petroleum Geochemistry.* Amsterdam: Elsevier, 1967.
Schurr, S. H., and B. C. Netschert, *Energy in the American Economy, 1850–1975. An Economic Study of its History and Prospects.* Baltimore: Johns Hopkins, 1960.
Williamson, I. A., *Coal Mining Geology.* New York: Oxford, 1967.

Index

GEOLOGIC TIME SCALE

AND SOME IMPORTANT DATES IN THE FORMATION OF MINERAL RESOURCES

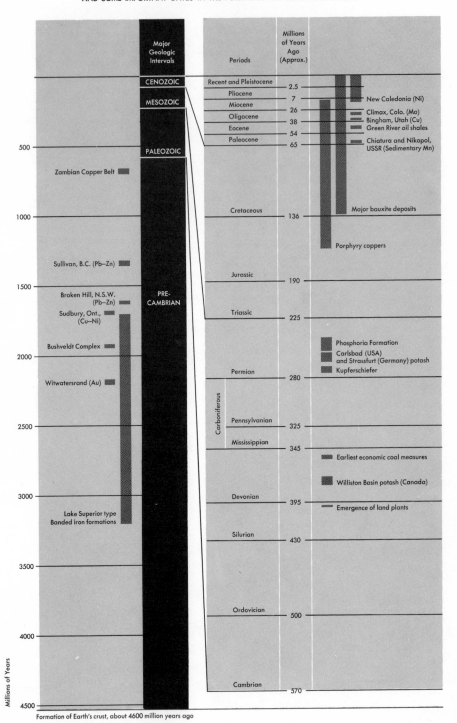

Formation of Earth's crust, about 4600 million years ago

Radiometric ages after Harland and others, 1964